3/1/08-3

Nuclear Weapons

This book is a history of nuclear weapons. From their initial theoretical development at the start of the twentieth century to the recent tests in North Korea, Jeremy Bernstein seeks to describe the basic science of nuclear weaponry at each point in the narrative. At the same time, he offers accounts and anecdotes of the personalities involved, many of whom he has known firsthand. Dr. Bernstein writes in response to what he sees as a widespread misunderstanding throughout the media and hence among the general public of the basic workings and potential impact of nuclear weaponry. For example, he points out that it has been nearly thirty years since anyone has even seen a nuclear detonation. Likewise, the Nagasaki bomb, primitive when compared to more modern devices, generated an explosion roughly the equivalent of eight thousand copies of the truck bomb used by Timothy McVeigh in Oklahoma City.

Jeremy Bernstein is Professor Emeritus of Physics at the Stevens Institute of Technology in Hoboken, New Jersey. He was a staff writer for the *New Yorker* from 1961 to 1995. He has written some fifty technical papers, three monographs, and twelve books, including *Albert Einstein*, which was nominated for a National Book Award; *Hitler's Uranium Club*; a biography of Robert Oppenheimer entitled *Oppenheimer: Portrait of an Enigma*; and most recently *Plutonium: A History of the World's Most Dangerous Element*.

Nuclear Weapons

WHAT YOU NEED TO KNOW

Jeremy Bernstein

CAMBRIDGE
UNIVERSITY PRESS

355.0217
B

CAMBRIDGE UNIVERSITY PRESS
Cambridge, New York, Melbourne, Madrid, Cape Town, Singapore, São Paulo, Delhi

Cambridge University Press
32 Avenue of the Americas, New York, NY 10013-2473, USA

www.cambridge.org
Information on this title: www.cambridge.org/9780521884082

First published 2008

Printed in the United States of America

A catalog record for this publication is available from the British Library.

Library of Congress Cataloging in Publication Data
Bernstein, Jeremy, 1929–
Nuclear weapons : what you need to know / Jeremy Bernstein.
p. cm.
Includes bibliographical references and index.
ISBN 978-0-521-88408-2 (hardback)
1. Nuclear weapons. I. Title.
U264.B453 2008
355.02'17–dc22 2007016625

ISBN 978-0-521-88408-2 hardback

CONTENTS

ACKNOWLEDGMENTS

I am indebted to a great many friends and colleagues for both criticism and helpful suggestions. Among the commentators were anonymous referees, whom I would like to thank here. The people I can thank personally, and I hope that I have not left anyone out, are Elihu Abrahams, Steve Adler, Norman Dombey, Sid Drell, Freeman Dyson, Richard Garwin, Roy Glauber, Sig Hecker, Walter Isaacson, Rainer Karlsch, Tom Lehrer, Norman Ramsey, Helmut Rechenberg, Cameron Reed, Carey Sublette, Richard Wilson, Herb York, and Peter Zimmerman. I also would like to thank from Cambridge University Press Simina Calin, Shari Chappell, Eric Crahan, Rufus Neal, and Laura Wilmot for their fine editorial work. As I was finishing this book I learned that my friend Francis Low had died. He and I shared an adventure in 1957 without which, as the reader will see, there would have been no book.

UNITS AND SIZES

These are the units and sizes needed for understanding fission and fusion. The "atomic mass unit" (amu) is commonly employed.

Mass Units

- 1 amu $= 1.66 \times 10^{-24}$ gram
- Mass of neutron $= 1.00867$ amu
- Mass of U-235 nucleus $= 234.9934 \sim 235$ amu $\sim 235 \times 1.66 \times 10^{-24}$ gram $\sim 3.9 \times 10^{-22}$ gram
- Number of U-235 nuclei in a kilogram $= 1,000 / 3.9 \times 10^{22} \sim 2.56 \times 10^{24}$
- 1 metric ton (tonne) $= 10^3$ kilograms

Energy Units

- 1 watt second $= 1$ joule $= 10^7$ ergs
- 1 electron volt $= 1.6 \times 10^{-19}$ joule
- 1 million electron volts (MeV) $= 1.6 \times 10^{-13}$ joule
- 1 kilo calorie $= 4.184 \times 10^3$ joules

It is common to express masses in terms of energy units using $E = mc^2$ with "c" the speed of light ($2.99792458 \times 10^{10}$ cm/sec). In these units the mass of the neutron is 939.573 MeV/c^2.

The average total energy in the uranium fission of one U-235 nucleus is about 200 MeV, which equals about 3.2×10^{-11} joule. The average energy of the fission of one Pu-239 nucleus is about 210 MeV. But the effective energy is about 175 MeV for U-235 or about 2.8×10^{-11} joule per gram.

- 1 kilogram of U-235 completely fissioned produces $2.8 \times 10^{-11} \times 2.56 \times 10^{24}$ joules $\sim 7.2 \times 10^{13}$ joules.
- 1 kilogram of TNT produces about 4.184×10^{6} joules $= 4.18 \times 10^{13}$ ergs.
- 1 metric tonne of TNT produces 4.18×10^{9} joules, so 1 kilotonne produces 4.18×10^{12} joules.

Thus it requires 72/4.2 kilotonnes, which is approximately 17 kilotonnes, or about 19 short kilotons, of TNT to produce an explosion equal to 1 kilogram of fissioned U-235. Timothy McVeigh's Ryder truck bomb gives us a practical unit: it had 2.5 tons of explosives. Therefore 1 kiloton equals about 400 Ryder trucks. McVeigh actually used ammonium nitrate, which has a comparable explosive power. The Nagasaki plutonium bomb had a yield of about 20 kilotons, which is approximately 8,000 Ryder trucks.

Hydrogen bombs obtain their yield by the fusion of light elements – primarily the isotopes of hydrogen of which there are three. The nucleus of ordinary hydrogen has one proton; heavy hydrogen – the deuteron (D) – has one proton and one neutron; and super heavy hydrogen – the triton (T) – has one proton and two neutrons. The most favorable fusion reaction used in hydrogen bombs is

$$D + T \rightarrow He + n + 17.6\,MeV.$$

In this expression "He" is the nucleus of ordinary helium with two neutrons and two protons and "n" is a neutron. The 17.6 MeV is the energy that is available because of the mass difference between the two sides of the expression. Most of this energy goes into the kinetic energy of the neutron. However, the charged alpha particle does most of the damage. In fission most of the energy goes into the fission fragments and much less to the neutrons, which have average energies of about 2 MeV. Uranium-238 has a neutron energy fission threshold of about 1 MeV. About three fourths of the neutrons produced in the fission of the uranium-235 in natural uranium have energies above this threshold, but only about one fourth are not slowed down below this threshold before the next fission. Hence there are not enough neutrons available above this threshold to fission the U-238 in natural uranium and produce a chain reaction. So natural uranium, which is more than 99 percent U-238 cannot be used to make a bomb. In a so-called hydrogen bomb a good deal of the yield is produced by subsequent fissions caused by the production of energetic neutrons in the fusion reactions.

The Hiroshima bomb, Little Boy, used 64.1 kilograms of about 89 percent enriched uranium and had a 1.4 percent efficiency, which would give about a 15-kiloton yield. The Nagasaki bomb used 6.2 kilograms of plutonium with a 17 percent efficiency, giving about a 20-kiloton yield. Hydrogen bombs produce a thousand times greater yield. In terms of Ryder trucks, the Nagasaki bomb was equivalent to eight thousand and a hydrogen bomb is equivalent to millions.

Introduction

Figure 1. Stanley Kubrick on the set of *Dr. Strangelove*. Courtesy the estate of Stanley Kubrick.

I N THE EARLY 1960S I SPENT PARTS OF A YEAR INTERVIEWING
Stanley Kubrick for what eventually became a *New Yorker* profile that was published in December 1966.[1] During my interviews he
was in the process of making *2001* but, inevitably, we talked about
Dr. Strangelove and the general subject of nuclear weapons. The
Cuban missile crisis had very recently occurred. Kubrick told me an
anecdote that I did not put in the profile. At the time of the crisis he
decided that he and his family would move to somewhere he deter-
ined would have the least fallout in case Washington or New York
was atom-bombed. He decided that Australia would do. Since he did
not fly – he thought that the odds against crashing were not small
enough – he had booked passage on a boat. He had also ordered 140-
odd Boy Scout camp trunks, which he was going to number and load
with the family goods. He then discovered that the bathroom of the
cabin that he was going to occupy with his wife was to be shared with
an adjoining cabin containing people he did not know. He canceled
the trip and decided to take his chances with the bomb.

[1] It was published in the December 11, 1966, issue under the title "How About
a Little Game?" Part of the interviews were tape-recorded and can be found in
the Stanley Kubrick Archives, Taschen, Berlin, 2005, as a CD.

What struck Kubrick, when we talked a few years after this incident, was how little interest there was among the general public about nuclear weapons. He thought that people had even less interest in them than they did in city government. He attributed this to the fact that, to most people, nuclear weapons are an abstraction. Almost no one had ever seen a nuclear explosion. The United States and Russia had stopped testing aboveground in 1962, so even the professionals who worked on the design of nuclear weapons in these countries no longer actually saw them explode. He said that people seemed to look on the absence of a nuclear conflict as they would money growing in a savings account. The longer there was no nuclear event, the safer we were. The whole premise of *Dr. Strangelove* was to show just how fragile all of this is.

The last aboveground test was conducted by the Chinese in 1980. So it has been close to thirty years since anyone has actually seen an atomic explosion. On the one hand we should be grateful and, on the other, concerned because people have so little understanding of what these weapons can do, or how they work. Most people seem to think that we are discussing explosions that are just a somewhat larger version of the car bombs in Baghdad. To put the matter in perspective, the Ryder truck that Timothy McVeigh used on April 17, 1995, to blow up the Murrah Building in Oklahoma City contained about five thousand pounds of high explosives. It killed 168 people. The *one* bomb that was dropped on Hiroshima on August, 6, 1945, had an explosive yield equivalent to about *thirty-two million* pounds of TNT. By comparison, on March 9 and 10, 1945, 334 B-29 bombers dropped two million pounds of incendiary bombs on Tokyo, killing perhaps 100,000 people. By December 15, 1945, it was estimated that 90,000[2] people had died in Hiroshima because of

---

[2] There is a good deal of uncertainty with respect to this number, but 90,000 seems to be the best conservative estimate.

the effects of *one* atomic bomb. The Oklahoma City bomb partially destroyed one building. The Hiroshima bomb destroyed 90 percent of the infrastructure of the entire city, and totally destroyed everything within a radius of one mile of where the bomb exploded. It may not be generally realized – I will explain more fully later – that this bomb was of a rather primitive design. Its design was considered to be so rudimentary that it had never been tested. The Hiroshima bombing was the test. The design resembled one that a country, or a group, that had acquired the explosive materials but had a limited technology would adopt. The bombs that make up our "portfolio" now typically have the equivalent explosive power of one *billion* pounds of TNT, and we have tested bombs that are a hundred times more powerful. As I will explain, their design requires a good deal of very sophisticated work. Some of these designs have been stolen and sold. This too I will discuss later.

For a while I thought that Kubrick might have been too pessimistic. In 1957, the United Nations established the International Atomic Energy Agency (IAEA) and, in 1968, a Nuclear Non-Proliferation Treaty (NPT) was drawn up. It was eventually signed by 190 countries. Israel, India, and Pakistan did not sign, and each of them constitutes a special case. For many years Israel maintained the fiction that it did not have nuclear weapons. But in 1985, a Moroccan-born Jew named Mordechai Vanunu, who had been fired from his job working at the plutonium-producing reactor in the Negev, defected and sold his story, with photographs, to the *London Sunday Times*. He was captured by the Mossad in Rome and, after serving seventeen years in prison, is living under tight security in Israel. We have gotten used to the idea that the Israelis have the bomb, and I doubt that many people in this country are kept up at night worrying about it. In 1974, the Indians tested their first nuclear weapon and, in 1998, the Pakistanis did the same. For a time it looked as if there might be a nuclear war over Kashmir. As horrible as this was to contemplate, considering the teeming cities of both countries and their limited capacity to respond to emergencies, it was to most of us very far away. I think most people in our country naïvely thought of such a nuclear exchange as *their* problem – *they* being the Indians and Pakistanis. In January 2003, North Korea withdrew from the non-proliferation treaty. Things began to look as if they were *our* problem. We had fought a war with North Korea and had troops stationed in the south. Furthermore, North Korea had a missile program that had as its goal the construction of intercontinental ballistic missiles capable of reaching the United States. It seemed very likely that the North Koreans had made at least some nuclear weapons and, indeed, on October 9, 2006, the North Koreans made an underground test of a low-yield nuclear device. Then there was Iran. Now things were getting uncomfortably close to home. Here was a country whose leaders not only appeared to be implacable enemies of ours, but who were also advocating the destruction of Israel. This was also a country that sponsored terrorism. That these people might build an atomic bomb was for many of us impossible to accept. Indeed, the media became replete with discussions of nuclear weapons. Terms like "fission," "enrichment," "centrifuge cascades," and "uranium isotopes" appeared on a daily basis.

What struck me about this was the appalling lack of understanding of any of it that manifested itself in most of these reports. One example, among very many, comes to mind. I was listening one Saturday morning to National Public Radio. An "expert" was being interviewed about the Iranian enrichment program. He referred to it as "distilling." The image came to mind of a large vat in which unenriched uranium was being boiled. The undesirable bits of uranium would emerge as a vapor, leaving behind just what you needed to make a bomb. The interviewer had absolutely no clue as to the total nonsense of this conception. It stood there unchallenged, adding to the growing budget of misinformation. Here is a more recent example. It comes from an article in the December 2006 *Atlantic* entitled "How to Get a Nuclear Bomb," written by William Lange-wiesche.[3] The article discusses the very important question of how terrorists might obtain nuclear weapons. Mr. Langewiesche who, whatever his virtues, is not a nuclear physicist, feels obligated to explain how a nuclear weapon works. He focuses on the Hiroshima bomb. He attempts to describe a "critical mass." Here is what he writes: "In relation to its surface area that mass was more than enough to achieve 'criticality' and allow for an uncontrollable chain of fission reactions, releasing further neutrons in a blossoming pro-cess of self-destruction." The late Wolfgang Pauli, who was both a great physicist and a scathing critic of nonsense, had a category of physics propositions that he said were not "even" wrong. They were so confused that they were all but devoid of meaning. This sentence is an example.

After having encountered a number of these it occurred to me that I could perform a service. Whereas I do not regard myself as a professional expert on nuclear weapons as compared to people like Freeman Dyson, Richard Garwin, Carey Sublette, Herbert York, or Pete Zimmerman, I have been interested in the subject for many years. I have taught nuclear physics and spent some time assisting Dyson when he was trying to design a nuclear bomb–powered space-ship. I am also one of the vanishing number of people still around who have actually seen the explosion of an atomic bomb – two, in

[3] *The Atlantic*, December 2006, pp. 80–98. The quote can be found on page 80.

fact. This occurred in the summer of 1957, when I was an intern at Los Alamos and went to observe two aboveground nuclear tests in Nevada. I will give my impressions of this experience later. In short, I decided to write a kind of nuclear weapons primer that would explain the basics and prepare people both to understand the news and to read the more complete accounts that can be found, for example, on the websites of Carey Sublette.

Here is how I have set about this task. I first give a history of the basic scientific discoveries, beginning with the discovery of the atomic nucleus just before the First World War and the discovery of the fission of uranium and plutonium just before the Second. I then describe how this science went to war. I also describe how this technology was stolen and the path that led from one country to the next. In the course of this I explain what enrichment is, why it is necessary, how it is done, and what a "critical mass" is. I also describe the hydrogen bomb and its history and how nuclear weapons work. I try to do this with an absolute minimum of technical jargon. I also mix the science with descriptions of the people who did it, something that I have always found fascinating. I hope the reader will be entertained – if this is a word that can be applied to a primer on nuclear weapons – and that, at the end, he or she will come away with a new understanding of the subject.

Let me make a remark in defense of my attempt to write about a subject about which so much has been written, including the two monumental books by Richard Rhodes, *The Making of the Atomic Bomb*, published in 1986,[4] and *Dark Sun*, published in 1995.[5] Whereas these books are extraordinary in their canvas, there are

[4] Simon and Schuster, New York.
[5] Simon and Schuster, New York.

things that have been left out or have come to light since their pub-lication. For example, nowhere in the books does Rhodes discuss the very bizarre chemistry of plutonium and the work of Cyril Smith, the British-born metallurgist who discovered how to make a workable alloy that could be used in making a weapon. Since these books were written the Russians have released some extraordinary new material that casts more light on the role of Klaus Fuchs and the hydrogen bomb. All of this is to say that this is a living subject about which the potential for discussion and discovery seems inexhaustible.

1. The Nucleus

Figure 2. Ernest Rutherford (right) with the "Talk Softly" sign above him. Photograph by C. E. Wynn-Williams, courtesy AIP Emilio Segré Visual Archives.

I AM GOING TO BEGIN MY DESCRIPTION OF THE ROAD THAT eventually led to the development of nuclear weapons from a starting place that may look a little whimsical – with the nineteenth-century German glassblower Heinrich Geissler. Geissler's father, Georg, was also a glassblower who made barometers and thermometers. Heinrich made glass tubes of very remarkable shapes. These tubes could be sealed and filled with various rarified gases or different liquids. When a high voltage was applied between the two ends of the tube there was a discharge. We would say that this arrangement caused an electric current to flow. Geissler was not, it seems, much interested in the science. He liked the fact that this discharge caused gas or liquid put in the tube to fluoresce, emitting bright colors. Geissler might be called the father of the neon sign. The first scientist to exploit these tubes was William Crookes in England. He was able to evacuate them so that whatever was being emitted from one end could move in a straight line. The motion was not being interrupted by collisions with molecules in the tube. He conjectured that the emitted objects were negatively charged particles. The heated wire that was emitting the particles became known as the "cathode." The receiver of the particles at the opposite end of the tube was called the "anode." Crookes discovered that if these "cathode rays" were allowed to impinge on certain minerals, they caused them to fluoresce and heat up. He thought that the "rays" might be negatively charged particles coming from the hot wire. This was verified by the French physicist Jean Perrin. He determined the sign of the charge by bending the rays in electric and magnetic fields. However, his experiments were not precise enough to determine any details such as the mass of the particles.

The great step forward was taken by the English physicist Joseph John Thomson, known ubiquitously as "J.J." J.J.'s father died when Thomson was 16, so it was difficult for him to go to university. But a mathematics teacher suggested that Thomson apply for a scholarship at Trinity College at Cambridge. He won the scholarship and never left the college – except for various visiting professorships – for the rest of his life. He was born in 1856 and died in 1940. Cambridge had a method for awarding honors to undergraduates. They took what was called the Tripos examination, one part of which could be devoted to some chosen field of study. The theoretically inclined scientist took the mathematical Tripos, which was notoriously difficult and required the ability to manipulate known formulae to solve well-posed problems very rapidly. The student who did best was known as the "Senior Wrangler." The next best was known as the "Second Wrangler," and the worst was known as the "Wooden Spoon." J.J. was Second Wrangler in his year. Curiously, the list of Second Wranglers produced scientists whose careers were very often much more distinguished than the Senior Wranglers. Rapidly solving well-posed problems does not necessarily imply scientific genius. One wonders how Einstein would have done. J.J.'s first interests were related to the theory of electricity and magnetism, which had been created by the Scottish physicist James Clerk Maxwell – the greatest physicist of the nineteenth century and another Second Wrangler. In 1896, J.J. visited Princeton, where he gave a series of lectures. It was upon his return to Cambridge that he began the series of experiments that would lead to his being awarded the 1906 Nobel Prize in Physics.

Thomson accepted the proposition that the cathode rays consisted of negatively charged particles. It had already been shown that their straight-line trajectories could be bent if they were sent through a magnetic field. What had not been shown was that sending them through an electric field would also bend the particle trajectories. It had been tried but no effect had been observed. Thomson reasoned that this was because the Crookes tubes had not been sufficiently evacuated and that the residual gas had interfered. The cathode ray particles collided with the gas molecules. He then improved the vacuum and showed that indeed the trajectories were bent in an electric field, proving that the particles had a negative electric charge. But what were they? To determine that, he had to find their mass.

Thomson had a very ingenious idea. The magnetic fields could be arranged so that they bent the particles in one direction while the electric field bent them in the opposite direction. But he could adjust the fields so that these two effects just compensated. Three things determined the magnitudes of the competing fields: the speed of the particles, their mass, and their charge. When he derived the equation for this it turned out that only the ratio of the charge to the mass entered, not their separate values. So, knowing the speed, which he did, he could determine the charge to mass ratio – e/m, where e is the charge and m the mass. But he had something to compare this ratio to. If one takes a hydrogen atom and disturbs it, it becomes electrified. To use the term of art, the hydrogen atom has become a hydrogen "ion." This ion has a positive charge. The ratio of its charge to its mass had been measured. What Thomson found was that for cathode rays this ratio was about 1,700 times larger than the hydrogen ion ratio. In his 1897 paper Thomson notes that this could be because the charge was larger or the mass smaller. He cast his vote for the smaller mass, which was confirmed when the charge was measured and shown to be identical in magnitude to the charge of the hydrogen ion. What Thomson had discovered was the "electron" – the charged particle that has the smallest mass in all of nature. The cathode rays were currents of electrons.

Because these electrons seemed to emerge from the atoms that made up the heated metal it seemed evident that they must have been inside these atoms to begin with. Thus, in 1904, Thomson proposed a model of the atom, which was often referred to at the time as the "plum pudding" atomic model. (I have looked at Thomson's 1904 paper and he does not seem to use this homey term.) It was well known that an ordinary nondisturbed atom is electrically neutral. But if you disturbed it, it could morph into a positively charged ion. Thus, in its nondisturbed electrically neutral form, it must, according to this picture, contain just enough electrons to neutralize the positive charge. If the positive charge is uniformly distributed, and the electrons placed here and there, then, if you are so-minded, they resemble plums stuck randomly in a pudding. Thomson tried to make all of this quantitative in ways that are now of little or no interest. However, there was one prediction that was significant. Suppose you could fire some sort of projectile into such an atom. Assuming that the projectile was a good deal more massive than the electrons, what would happen? It was reasoned that the projectile might suffer a series of tiny deflections when it encountered the light electrons. But it would eventually emerge from the atom in more or less the same direction from which it had entered. The probability of a very large deflection, according to this model, was infinitesimal. The whole operation was like shooting bullets through fog. Enter Ernest Rutherford.

Rutherford was born on August 30, 1871, in Nelson, New Zealand. His father, who was a wheelwright – a repairer of wheels for carts or wagons – had emigrated from Scotland. Rutherford was educated in local schools, matriculating at the University of New Zealand in Auckland. He was an outstanding student and, in 1894, won a scholarship to Trinity College to work with Thomson. In 1898, a position opened up at McGill University in Montreal. There he encountered a somewhat younger English chemist named Frederick Soddy. Soddy became the first of an astonishing list of Rutherford students and collaborators who went on to win the Nobel Prize. Soddy won his in chemistry in 1921. Rutherford had won his in chemistry in 1908. Rutherford was very surprised to win the chemistry prize rather than the physics prize, but he graciously accepted it. He and Soddy went to work on the fairly new subject of radioactivity. In a series of very important papers they showed that radioactivity is something that is intrinsic to atoms. You can take a collection of radioactive atoms, put them in a fire, paint them blue, or leave them out in the rain and they will continue emitting the same particles at the same rate with the same energies. Rutherford recognized that there were three kinds of radioactive emission, which he called alpha, beta, and gamma: α, β, and γ. The gamma rays were electrically neutral radiation, even more penetrating than X-rays. After Einstein proposed the particle nature of radiation in 1905, it was understood that the energy of such radiation "quanta" was proportional to their frequency of vibration. Thus the gamma rays were composed of quanta of higher frequency than X-rays. The beta particles were soon identified as Thomson's electrons. That left the alpha particles. Rutherford conjectured that these were ionized helium atoms. This was based on the fact that they had a positive charge that was twice that of ionized hydrogen and their mass was four times as large, fitting what was then known about helium atoms. Moreover, on Earth, the very rare element helium was often present around radioactive elements like uranium or radium, suggesting that some chain of decays was producing helium. All of this was confirmed when Rutherford later showed that the atomic alpha particles had the same pattern of spectral lines – colors of light – as helium atoms.

In 1907, Rutherford returned to England to accept a professorship at the University of Manchester. During the next few years he did probably the most important work of his very productive career. One thing was always true of Rutherford: his self-confidence was boundless. He knew just how good he was and he presented his ideas in a booming voice. That someone would take a picture of him in his laboratory under a sign that reads "Talk Softly Please" is not an accident. But Rutherford had a wonderful sense of the worth of the young researchers who came to work with him. For example, Niels Bohr, who had gotten his Ph.D. at Copenhagen University in 1911, came to England to do postdoctoral work that year. First he went to Cambridge to work with Thomson. He did not find that very satisfying so he then went to Manchester to work with Rutherford. Bohr was shy and very unassuming, the mirror image of Rutherford. But Rutherford saw in Bohr the genius that led Bohr, upon his return from England in 1913, to formulate the first quantum theory of the atom. It had been inspired by the discovery, which I am about to explain, made by Rutherford.

One of Rutherford's very first assistants at Manchester was a young German physicist named Hans Geiger. Geiger later went on to invent the radiation detector that bears his name. Still later he became a loathsome Nazi who turned in his Jewish colleagues. Geiger came to work with Rutherford in 1907 in the halcyon days before the First World War. Even when he was still in Canada, Rutherford had done experiments with his beloved alpha particles that showed that when they passed through matter they were apparently only slightly deflected, as the plum pudding model would suggest. He proposed that Geiger look into this more carefully. To this end, alpha particles generated by the decay of radium were made to impinge on ultra-thin foils of gold. They were so thin that it took only about four hundred atoms to span a width. This was important because then one could have some confidence that what one was observing was the scattering from individual atoms and not some collective effect. In 1909, Geiger came to Rutherford with the request that an undergraduate named Ernest Marsden be brought into the collaboration. What then happened was marvelously described in a much later lecture by Rutherford.

I had observed the scattering of alpha-particles, and Dr. Geiger in my laboratory had examined it in detail. He found in thin pieces of heavy metal the scattering was usually small, of the order of one degree. One day Geiger came to me and said, "Don't you think that young Marsden, whom I am training in radioactive methods, ought to begin a small research?" Now I had thought that too, so I said, "Why not let him see if any alpha-particles can be scattered through a large angle?" I may tell you in confidence that I did not believe they would be, since we knew the alpha-particle was a very fast massive particle with a great deal of energy, and you could show that if the scattering was due to the accumulated effect of a number of small scatterings, the chance of an alpha-particle's being scattered backward was very small. Then I remember two or three days later Geiger coming to me in great excitement and saying "We have been able to get some alpha-particles coming backward…" It was quite the most incredible event that ever happened to me in my life. It was almost as incredible as if you fired a 15-inch shell at a piece of tissue paper and it came back and hit you.[1]

The alpha particles had hit something hard inside the gold atom. It was the atomic nucleus. Its discovery changed everything in physics, and not long after its discovery, the First World War changed everything in the lives of people. Geiger and Marsden both served on the Western front but on opposite sides of the line.

[1] This is quoted in a very instructive website created by Michael Fowler: http://Galileo.phys.virginia.edu/classes/252/Rutherford_Scattering.

2. Neutrons

Figure 3. Enrico Fermi. University of Chicago, courtesy AIP Emilio Segré Visual Archives.

I T TOOK RUTHERFORD SOME TIME TO ASSIMILATE HIS OWN discovery. It turned the atom inside out. In the plum pudding model the electrons were inside the atom but in his new model they were on the outside. Because they were presumably moving around, the laws of electricity and magnetism implied that they would radiate. But a radiating electron would lose energy and fall into the nucleus. What kept the whole thing stable? This was the problem that Niels Bohr addressed in 1913, when he introduced his quantum theory of electron orbits. In 1911 Rutherford published the theory of the nuclear discovery experiment, or, at least, his theory. This is a classical Newtonian theory in which the alpha particle moves in an orbit that is a hyperbola. He got this simple result because he assumed that the nucleus was so small compared to the size of the atom that one could think of the positive charge of the nucleus as if it were concentrated at a point. What he derived was the relative chance that the alpha particle could scatter at some given angle. When he compared his result with the experiment of Geiger and Marsden there was agreement. Here is a curious aside: if one uses Rutherford's point charge assumption in the modern quantum theory, one gets the same result he did. His 1911 formula is still with us. From the agreement of his result with the experiment, which assumed that the alpha particle deflection was caused by collision with a single nucleus, he was able to estimate the size of a gold nucleus: a radius of less than 10^{-11} centimeter. In a later experiment he found it to be about 3×10^{-12} centimeter. The modern value is a little more than twice as big. Nonetheless this radius is about ten thousand times smaller than the size of the atom, which is determined by the distribution of the electrons. To give some idea of the scale, the radius of the Earth is about 6,400 kilometers whereas the average distance to the Sun is about 150 million kilometers. If the gold nucleus were the size of the Earth, the electrons would be not terribly far from the Sun.

War broke out in 1914, and for its duration Rutherford concentrated mainly on war work. In 1919 he accepted the position as Thomson's successor at Cambridge, and there he stayed for the rest of his professional life. He died in 1937. Once he got to Cambridge he began focusing on the question of what the nucleus was made of. Clearly, he reasoned, it must contain positively charged particles. The simplest of the nuclei – hydrogen – has one of them, which Rutherford named the "proton." But charge does not follow mass. For example, the helium nucleus has a charge of two proton charges but a mass of roughly four protons. There had to be noncharged particles in the nucleus of about the same mass as the proton. Rutherford made a reasonable – but incorrect – guess as to what they were. He conjectured that they were protons and electrons bound together. Now enter James Chadwick.[1] Chadwick was born in Chesire, England, in 1891. He died in 1974. During

---

[1] For an excellent description of his life and work with insights about Rutherford, see *The Neutron and the Bomb*, by Andrew Brown, Oxford University Press, New York, 1997.

the Second World War he had led the British part of the effort to make the atomic bomb. In 1908, he got a scholarship to the University of Manchester, where he made a very favorable impression on Rutherford, who kept him on as a graduate student. In 1913, he won a fellowship that enabled him to continue his studies at a continental university and, presumably following the advice of Rutherford, he chose to study with Geiger in Berlin. Chadwick had fond things to say of Geiger who, among other things, advised him to get out of Germany rapidly once the war began. Chadwick did not take this advice soon enough and found himself interned with a variety of other British prisoners of war in the former stables of the race track at Ruhleben. There he remained for the duration. But, despite the conditions, he managed to keep doing science.. The detainees organized a "Science Circle" at which Chadwick gave lectures about radioactivity. He also did a few pretty primitive experiments – one of them involving the radioactivity in a brand of toothpaste – with a fellow detainee named Charles Drummond Ellis, with whom he would later collaborate at Cambridge. Indeed, when the war ended Chadwick went to Cambridge to continue his work with Rutherford toward a Ph.D. He then became Rutherford's assistant. But it was not until the early 1930s that Chadwick did the experiments that led to the discovery of the neutron.

Like most discoveries in physics – Einstein's being an exception – Chadwick's was built on the work of others. Polonium, one of the radioactive elements that Madame Curie had found and named after her native Poland, was a strong emitter of energetic alpha particles. These could be used to bombard various elements. This had been done by Chadwick and Rutherford and, in 1928, by the German physicist Walther Bothe and his student Herbert Becker. During the Second World War, Bothe spent much of his time trying to build a cyclotron. He was not a member of the Nazi Party so his efforts never received high priority. He did not complete building it until practically the end of the war, and then he could not operate it because of the bombing. He also did an experiment, which I will discuss later, that more or less doomed the German atomic bomb program. In 1928 Bothe found that alpha particles, when they impinged on elements like aluminum, boron, and magnesium, induced these elements to produce gamma radiation. He tried this on beryllium and got results that he could not explain. This experiment was repeated by Irène Curie, Marie's daughter, and her husband, Frédéric Joliot. They concluded that the radiation from the beryllium was gamma radiation but of a higher energy than had ever been previously observed.

Chadwick decided that they must be wrong. Among other things gamma radiation would be emitted in all directions, whereas this radiation was emitted primarily in the direction of the incident alpha particle, as if it had taken up the momentum of the alpha. Chadwick was sure that the alpha had been absorbed by the beryllium nucleus and had transformed it into a nucleus of carbon, ejecting one of Rutherford's neutrons, which took up the momentum of the alpha. The residual carbon nucleus was too heavy to move much. The neutron is electrically neutral, but when it goes through a gas it collides with some of the atoms, knocking out some of their atomic electrons. This leaves a trail of positively charged atoms – ions – and these can be detected. One can even photograph the trail of ions with a suitable detector. Chadwick was certain that gamma radiation would never have been able to do this. The gas atoms were being hit by something substantial.

He staked his claim in a brief letter published in the February 27, 1932, issue of the British journal *Nature*. He followed this a few months later with a long paper in the *Proceedings of the Royal Society*.[2] In this paper he gives his argument as to why he is convinced that the "neutron" he has found is, indeed, Rutherford's electron and proton stuck together. The elements of this argument will be very important to us when we discuss the fission and fusion reactions in nuclear weapons. Suppose that Rutherford was right. Then to split the neutron back into its component parts – a proton and an electron – we would have to supply energy to induce the breakup. The amount of energy we have to supply is called the "binding energy." But, in 1905, Einstein demonstrated in his papers on relativity that all energy corresponds to an equivalent mass according to the equation $E = mc^2$, where "c" is the speed of light. In this context this means that the bound electron and proton must be *less* massive than the free electron and proton. The difference in mass is, by Einstein's formula, related to the binding energy. In short, the whole must be less massive than the sum of the parts. Chadwick had measurements of the masses of the proton, electron, and neutron, and they seemed to show that the neutron was less massive than the sum of the proton and electron masses, which would make the case for Rutherford's bound electron-proton hypothesis. He summarizes by saying, "It is, of course, possible to suppose that the neutron may be an elementary particle. This view has little to recommend it at present, except the possibility of explaining the statistics of N^{14}."[3]

The last part of this remark refers to a quantum mechanical argument that Chadwick was dismissing out of hand. In fact, the argument was right. The measurements of the masses Chadwick used were wrong and, in fact, the neutron *is* an elementary particle – at

---

[2] Chadwick, *Proc. Roy. Soc.*, **A, 136**, 692–708 (1932).
[3] Chadwick, op. cit., p. 706.

least as elementary as the proton – whose mass is greater than the sum of the proton and electron masses. Because the neutron is more massive, it can decay into a proton, an electron, and a neutrino, a ghostly, uncharged particle with a tiny mass. Chadwick was awarded the Nobel Prize in Physics for 1935, three years after his discovery. But enough time had gone by, and enough work had been done in the interim, so that in his Nobel lecture he reversed himself completely. He acknowledged the correctness of the quantum mechanical argument. He quoted new and more accurate measurements of the masses and discussed the instability of the free neutron. That is how science works at its best. It corrects itself as it goes along. The neutron was here to stay as a particle in its own right.

In Greek "isos" means "equal" or "same" and "topos" can mean "place." This was contracted into the word "isotope" – the "same place." The "place" here refers to the position of an element in the periodic table of elements. An element's place in the periodic table is determined by the charge of its nucleus – the number of protons. Thus hydrogen is in the first place and oxygen, with its eight protons, is in the eighth place. The term "isotope" entered science in a 1913 Chemical Society annual report written by Frederick Soddy, the chemist who had worked with Rutherford in Montreal. He left Canada in 1902 and in 1913 was at the University of Glasgow. In the meantime he had gotten married, and a friend of his in-laws, Dr. Margaret Todd, suggested the name "isotope" to describe what Soddy had been discovering. I am unable to supply the circumstances of Dr. Todd's suggestion. In any event, what Soddy found is that elements like radium and thorium seemed to manifest themselves in forms that differed by their radioactivity. For a while there was a good deal of confusion, as is indicated by the names given to these manifestations. To give two examples, there was something called "mesothorium-I" that was chemically indistinguishable from radium but differed from ordinary radium because of its type of radioactivity. There was also something that was called "thorium-A" that was chemically indistinguishable from polonium but again differed from ordinary polonium. Soddy proposed to call thorium-A, for example, an "isotope" of polonium because it had the same chemical properties as polonium and hence would be put in the same place in the periodic table. In the context of the plum pudding model this sort of thing was incomprehensible, but in terms of the Rutherford model it became totally clear.

In this model, the electrons distant from the nucleus determine the chemistry of the atom, which involves the activity of the electrons, while the distant heavy nuclear particles are far-off spectators. But the number of electrons in an electrically neutral atom is equal to the number of protons. What fixes an element's place in the periodic table is the number of protons in its nucleus, which determines the number of electrons and hence the chemistry. But there can be different nuclei with the same number of protons and different numbers of neutrons. These variations are what constitute the isotopes. In the preceding examples, "mesothorium-I" was actually an isotope of radium with 88 protons and 136 neutrons, whereas a common isotope of radium has 88 protons and 134 neutrons. Likewise, "thorium-A," which is an isotope of polonium, has 84 protons and 132 neutrons. All thirty-four isotopes of polonium have 84 protons and various numbers of neutrons. To designate the isotopes, it is very common to add the number of neutrons and protons and then write that number in the upper left of the chemical symbol. This sum is often referred to as the "atomic weight." So we would write the isotope of polonium formerly know as thorium-A as ^{216}Po. The number 216 is the atomic weight of the isotope. To aid in remembering the number of protons, which determines its place in the periodic table, that number is usually written in the lower left. Thus, the full symbol would read $^{216}_{84}$Po. The reader might amuse him- or herself by doing the same thing with radium, whose symbol is Ra.

If you have an element like polonium with thirty-four isotopes, you must do something about keeping them straight. Soddy's 1922 Nobel lecture – he won the prize in 1921, the same year as Einstein, who gave his lecture the following year – is somewhat confusing because he was still convinced that the neutron was a composite of a proton and electron so he referred to positive and negative charges in the nucleus. This was all cleared up once Chadwick discovered the neutron and it became clear that it was a particle its own right.

The isotopes discovered by Soddy and others were all radioactive. This is not surprising because they were detected by their different radioactivities. J. J. Thomson raised the question of whether there are stable isotopes. This required a different kind of detection. How to do this was suggested by the technique that Thomson had used to determine the electron mass. One would try to bend the orbits of the charged ions of elements suspected to have isotopes. Thomson had an assistant at Cambridge named Francis Aston. Aston acquired the job of carrying this out. He was born in 1877 in Birmingham and was educated at Mason College, which later became the University of Birmingham. But then he left academia and worked for three years as a chemist in a brewery, after which he returned to the university, where he worked on Crookes tubes. In 1909, he became Thomson's assistant. He began sending various gases through the Crookes tubes and bending the ions with electric and magnetic fields as Thomson had done. In 1912, when he tried this with neon, something odd occurred. In all the previous experiments ions had followed a single orbit – a parabola. But for neon there were two distinct orbits. The ions were collected on a photographic plate so that the intensity corresponding to the two orbits was registered. In the case of neon there was a strong intensity for mass number 20 and a weak one for mass number 22. Aston had discovered two isotopes of neon. There are three stable neon isotopes, but the one with mass number 21 is too rare for Aston to have observed it.

Aston's work was interrupted by the First World War, during which he worked on properties of the fabrics used in airplanes. But, in 1919, he was back at Cambridge working on isotope separation. He had decided that the Crookes tube method with the photographic plates was not precise enough and that something new was needed. To find it he used an analogy with light. Light can split into beams of different colors by passing it through a prism. Aston wanted to split a beam of atoms into different beams that reflected the masses of the various isotopes. In essence it worked as follows, although Aston's actual device was more complex. The element to be analyzed in gaseous form is introduced into a vacuum chamber. It is then ionized by being bombarded with a stream of electrons that knock the electrons off the atoms. The now positively charged ions are accelerated by an electric field. The heavier isotopes are more difficult to accelerate and hence acquire a slower speed. The groups of isotopes traveling at various speeds are then subjected to a magnetic field. This bends the orbits into semicircles. The heavier the isotope is, the less it is bent and the larger the radius of its circular orbit. This means that the original gas has now been split into isotopes moving along different semicircles. These different beams can be detected on photographic plates, where they show up as parallel lines. The device used to accomplish this is called a mass spectrometer. Figure 4 shows how the information produced by a mass spectrometer looks. The heights of the different lines measure the

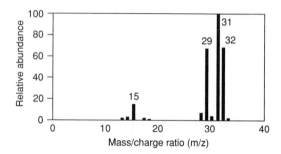

Figure 4. The relative abundances of isotopes produced in a mass spectrometer.

relative abundance of the isotopes. I go into this detail because electromagnetic separation was one of the methods used in the Second World War for separating uranium isotopes. In 1922 Aston won the Nobel Prize in Chemistry for this work.

I MET ENRICO FERMI ONLY ONCE. IT WAS THE SPRING OF 1954, AND he had come to Harvard to give a series of so-called Loeb Lectures. I was a graduate student finishing my thesis. Roy Glauber, then a very young assistant professor – he won the Nobel Prize in Physics for 2005 – had organized a kind of seminar for advanced graduate students and postdoctorals in theoretical physics. It was an informal sort of thing in which you could try out ideas and learn from one another. Somehow he persuaded Fermi to give a talk to us. Fermi chose a topic in elementary quantum mechanics. I remember the topic but I forget what insight Fermi had. It involved some approximations. We all sat listening to this in respectful silence except for one of us – a very brilliant fellow who went on to have a fine career – who announced that he did not believe in Fermi's approximations. Fermi then gave a second lecture with fewer approximations. Our fellow said once again that he did not believe in all the approximations. Fermi then gave a third lecture with many fewer approximations. Our fellow left well enough alone. I thought that Fermi would go on all afternoon. There was no depth at which he did not know the subject.

Fermi's scientific genius was recognized very early. He was born in Rome on September 28, 1901. He took his Ph.D. when he was 21 and was a full professor at the University of Rome when he was 26. What made Fermi unique was his mastery of all branches of physics, both experimental and theoretical. No one else had this mastery, and it is likely no one will ever again because of the complexity of the subject. To give an example, the fact that the neutron was a particle that did not contain an electron posed a problem. As I have mentioned, the neutron is unstable and decays into a proton, an electron, and a neutrino. In the Rutherford composite model of the neutron, the answer to the question of where the electron came from was clear. It was there all the time. But if the neutron was elementary, where did the electron come from? Fermi showed that, in the quantum theory, it was created at the time of decay. Although this idea had been used in the theory of gamma decay, it had never been applied to a situation in which a massive charged particle could be created. This was something quite novel: the notion that such particles could be created and annihilated. A vast amount of work has gone into the theory of beta decay in the past seventy years, and Fermi's theory is at the heart of it. But that same year he began an experimental program that, in 1938, won him the Nobel Prize. He collected the prize in Stockholm and then went on to the United States with his family. His wife was Jewish, and it had become too dangerous to remain in Italy.

Prior to Fermi's work, the probe of choice of the nucleus was the alpha particle. But there was a problem. The alpha particle has a positive electric charge because of the two protons in its nucleus. Nuclei all have positive charges. But like charges repel. Thus when the alpha particle gets close to a nucleus it is repelled. The alpha particle always has a distance of closest approach. But the neutron is electrically neutral. It can penetrate right into the nucleus, opening up all sorts of possibilities for creating new isotopes or even new elements. Many people understood this concept, but Fermi's group in Rome was the first to actually take advantage of it. By 1934, he had built up a small but very powerful group, some of whom went on to win Nobel Prizes in their own right. For the fun of it, they gave themselves ecclesiastical names. Fermi was, of course, the Pope, and Orso Corbino was known as the Heavenly Father. Corbino ran the physics department and was a state senator, which meant that he kept the group well financed. They began to subject one element after another to neutrons to see what would happen. They even borrowed a gold ingot. But, in the fall of 1934, they began to notice some very odd effects. The intensity of the reactions appeared to depend on what kind of a surface the apparatus was sitting on. When it was on a wooden surface the intensity increased dramatically. For reasons that Fermi could never explain, he decided to put a piece of paraffin in front of the target. There was a dramatic increase in the intensity. When they tried other materials, nothing happened. This experiment took place in the morning, and at lunchtime Fermi went home for a siesta. By the time he returned that afternoon, he had figured out what had happened and had invented a new branch of physics.

Paraffin is a hydrocarbon consisting of hydrogen and carbon atoms. When a neutron enters the paraffin, it begins colliding with these atoms. In each collision these atoms absorb some of the neutron's momentum. After several collisions the neutron is slowed down to something like the speed of sound. These slow neutrons spend more time within a nucleus of, say, gold. So there is a greater chance of an interaction. This in essence explains the increase in the intensity of these reactions when the neutrons are slowed down, although one needs the quantum theory for the details. Fermi had discovered what the reactor scientists later called a "moderator" – something that slows down neutrons in a reactor. Indeed, when in 1942 Fermi designed the first reactor, it used a highly purified form of graphite – again carbon – as its moderator.

Having worked up the periodic table, the heaviest known element yet to be tried was uranium. But Fermi and his group had an idée fixe. They were certain that the neutron would penetrate the uranium nucleus and make a new uranium isotope. This isotope would be radioactive and when it decayed by beta decay, for example, a neutron would be converted into a proton. But this would then be part of the nucleus of a wholly new element beyond uranium in the periodic table. In short, they were convinced that they were going to discover "transuranic" elements. Indeed, when the new kinds of radioactivity were observed they were sure that that was what they had discovered. So was the Nobel committee. When Fermi was presented to the king of Sweden in the Nobel ceremony of 1938, the presenter noted that

The general pattern that Fermi has found to be the rule when heavy substances are subjected to irradiation by neutrons, took on special interest when applied by him to the last element in the series of elements, viz. uranium, which has rank number 92. Following this process, the first product of disintegration should be an element with 93 positive electric charges and a new element would thus have been found, lying outside the old series. Fermi's researches on uranium made it most probable that a series of new elements could be found, which exist beyond the element with rank number 92. Fermi even succeeded in producing two new elements, 93 and 94 in rank number. These new elements he called Ausenium and Hesperium.[4]

Ausenium and hesperium are nice names. However, these elements are now called neptunium and plutonium, and Fermi did not discover them. They were discovered a few years later using the cyclotrons at Berkeley. What Fermi had actually discovered, although for reasons I will explain he did not realize it then, was nuclear fission. In fact, at almost the same time that Fermi was getting his Nobel, at least in part for something he did not discover, Otto Hahn and Fritz Strassmann in Germany were actually discovering nuclear fission, and the race for the bomb, in which Fermi would play a major role, began.

[4] http://nobelprize.org/nobel_prizes/physics/1938/press.html.

3. Fissions

Figure 5. Group photo taken in Berlin in 1920: Lise Meitner is in the center with Bohr to her right and Otto Hahn directly behind her. The Niels Bohr Archive, Copenhagen.

Oh what idiots we have been! Oh but this is wonderful. This is just as it must be![1]

> – Niels Bohr; upon being informed (by Otto Frisch) of Hahn and
> Strassman's "strange" results (January 3, 1939)

THE QUOTATION ABOVE IS WHAT NIELS BOHR SAID TO OTTO Frisch when Frisch returned to Copenhagen after spending his Christmas vacation with his aunt Lise Meitner in Sweden. The two of them had just realized that the experiments – which I will describe – of Fritz Strassmann and Otto Hahn revealed the fissioning of the uranium nucleus, and this was Bohr's reaction. I have heard a number of nuclear physicists who were active at that time make similar comments. Robert Serber, for example, who played a crucial role at Los Alamos, once said to me, "How could we have been so stupid?" – "stupid" not to have thought of fission. Another person who reacted the same way was Hans Bethe. In 1937, the year before fission was discovered, he published three monumental review articles on nuclear physics in the *Reviews of Modern Physics*. They became known as "Bethe's Bible," and they contained everything that was known about the subject. In retrospect Bethe simply could not understand why he had missed predicting fission. The one person I never heard, or read about, saying anything like this

---

[1] This is quoted in *Erindringer om Niels Bohr*, edited by Stefan Rozental, Gylendal, Copenhagen, 1985.

was Fermi, even though his neutron experiments should have shown nuclear fission. Why didn't they? The experimenters expected that they were going to produce transuranic elements. They were sure that these would be very radioactive. Indeed they worried that their activity might swamp their detectors. They decided to cover their target sample of uranium with an aluminum foil, which would block some of the undesirable radiation. The trouble was that it also blocked the fission fragments that would have produced a pulse in their detector. Given Fermi's genius, I am sure that had these fission fragments not been blocked, he would have figured out what was happening. Uranium fission would have been discovered in Fascist Italy in 1934 rather than Nazi Germany in 1938. What would this have meant for the war? One can only imagine.

There was one person who did not believe Fermi's result and said so in no uncertain terms – in print. This was the German chemist Ida Noddack. In 1925, she and her husband, Walter, had identified a new element that they first called "rhenus" after the Rhine. It is now called "rhenium." Noddack had several objections to the Rome group's work. In the first place they did not cite her work. In addition, they communicated what they had done to the newspapers. But her scientific objections were the most serious. As I have mentioned, what the Fermi group did was look for radioactive decays that had been produced by neutrons incident on uranium. They found a pattern that did not fit any of the elements close to uranium in the periodic table. They were sure that whatever had taken place could not have produced any element far from uranium on the periodic table. Noddack argued that this assumption was unjustified. What they should have done was examine elements in the middle of the periodic table to see if anything fit. She said that what might well have happened was that the neutron split the uranium nucleus. She was simply not taken seriously. This was in part, I think, because some of her chemical discoveries were suspect. But the real reason was that she did not supply a mechanism. How could a slow-moving neutron cause a massive uranium nucleus to fission? I am sure that this very reasonable question is what stopped the other nuclear physicists of this era from predicting fission.

The ultimate discovery of fission was intertwined with politics – in particular the rise of the Nazis. Of the four people involved, Otto Hahn, Fritz Strassmann, Lise Meitner, and Otto Frisch, two of them – Frisch and Meitner – had been born Jewish. Meitner was born in Vienna in 1878 to a prosperous middle-class family.[2] Her father, a lawyer, believed in the education of women, which meant that she had to be privately tutored in science and mathematics. She entered the University of Vienna at the age of 23 and had the good fortune to find a great teacher, Ludwig Boltzmann, who also believed in the education of women. In 1906, Meitner got her Ph.D. in physics, the second woman in Austria to do so, and then realized that if she stayed in Vienna her only future was as a schoolteacher. So in 1907, she went to Berlin. There she met Hahn, who was four months younger than she. Hahn was one of those people who, due to their ability and charm, seem to float through life. He was a good but not particularly dedicated student. After his Ph.D. in 1904 he went to Cambridge to do postdoctoral work. He switched his field to radio chemistry and discovered what turned out to be a new iso-tope of thorium. Because the notion of isotopes had not yet been invented he thought he had discovered a new element. Then he went to Montreal to work with Rutherford. In 1906, he was offered

[2] For an excellent biography of Meitner, see *Lise Meitner, A Life in Physics*, by Ruth Sime, University of California Press, Berkeley, 1996.

a position in the laboratory of a very distinguished German chemist named Emile Fischer. Meitner also found herself in Fischer's laboratory, but through the back door. Her professor, Heinrich Rubens, persuaded Fischer to give Meitner a small space in the basement of his laboratory. There was no lady's room, so Meitner had to use the facilities in a nearby restaurant. She was also under orders not to enter any laboratory in which there were male students. It goes without saying that she was not paid. But soon she and Hahn began collaborating, she as a physicist and he as a chemist.

They had two periods of collaboration: from the time they met until the First World War, in which Hahn worked on poison gas – his response was always that someone had to do it – and then from 1934 to just before the discovery of fission. In 1912, an entity called the Kaiser Wilhelm Institute for Chemistry was created. It was independent of the universities and was a national laboratory designed to attract talented young scientists. Hahn got a job there with the title of Professor and a good salary. In 1917, Meitner got a position there, originally as a "guest," although she was asked to organize the physics department. She received a stipend that was considerably less than Hahn's. While Hahn was still busy with his war work she finished writing a joint paper on a new isotope, for which he got essentially all the credit, although she had done most of the work. She then stopped collaborating with him until 1934. By that time the Nazis were in power. Meitner temporarily staved off the inevitable because of three things, of which the last was the most important. In 1908, Meitner had converted to Christianity, which would not have exempted her from the racial laws in the universities if she had been in a university. But her institute was exempt from those laws, although it came under pressure because it had hired too many non-Aryans. Most importantly, she was an Austrian citizen, which again exempted her from the Nazi racial laws, until the spring of 1938, when Hitler annexed Austria. Then she had to leave Germany, barely escaping with her life to Holland.

In the 1930s Meitner was, as far as scientific discoveries were concerned, in the "loop." By that time she was well known and her laboratory had frequent visitors. After his discovery of the neutron, Chadwick visited her and came home with some loaves of pumpernickel she had gotten for him. One of Fermi's group, Franco Rasetti, came for a visit and went back home with some designs for new detectors of radioactivity. She received papers from Joliot and his wife Irène Curie. They had produced artificial isotopes, using alpha particles, for which they received the 1935 Nobel Prize in Chemistry. And she received papers from Fermi on his neutron experiments.

She decided that she was going to begin her own neutron experiments. She needed a chemist and she tried to enlist Hahn. At first he was not interested, but eventually he joined her. They were also joined by a younger chemist named Fritz Strassmann. Strassmann was born in 1902, in Boppard, Germany. He got his Ph.D. in chemistry from the Technical University of Hannover in 1929. Afterward he came to the Kaiser Wilhelm Institute to work for a year with Hahn and, liking it there, he stayed on with a small stipend. Then the Nazis came to power. Strassmann was, unlike Hahn, a courageous anti-Nazi. Hahn might best be described as a "non-Nazi." He did not join the party, but he did not resist either. Strassmann also refused to join the party and lost the chance for a lucrative job in industry. Later, he and his wife hid a Jewish musician named Andrea Wolfenstein in their apartment. If they had been discovered, all of them, including Strassmann's infant son, would have been executed. As it was, they were practically starving. All Hahn could pay him was half the normal stipend. After the war Strassmann was named a "Righteous Gentile" by Yad Vashem in Jerusalem, and a tree was planted in his honor.

Meitner's idea was to study the transuranics that Fermi had allegedly found. She had discovered independently that slow neutrons enhanced the reaction rates, so she used a less energetic neutron source. She was sure that the neutrons would produce transuranics and that these would have the same chemistry as the so-called transition elements such as rhenium. Rhenium lies just above where element 93 should be in the periodic table, so this seemed like a reasonable assumption. It happened to be wrong. By 1936, the group was confident that it had evidence for the transuranics and said so in print. Some things didn't quite fit, but they were so sure of their basic premise that they brushed the objections under the rug. The last paper the three of them published together was in 1938. Then the roof fell in on Meitner.

Until 1938, she simply chose to ignore the warning signals that were all around her. Hahn had been asked to remove the non-Aryans from his laboratory and, as he was Meitner's nominal director, he asked her to leave, which she refused to do. She had received a feeler from Swarthmore that she did not explore because, as an undergraduate institution, it did not offer the proper laboratory facilities. I have often asked myself what would have happened had she emigrated to the United States before 1938. She certainly would not have participated in the discovery of fission. As we shall see, that depended on her being in close communication with Hahn. Indeed, she would have been, in truth, a rather minor footnote in the history of twentieth-century physics. And who would have discovered fission? Hahn would have made the experiments I am going to describe, but there is no evidence that he would have correctly interpreted them. Some other physicist – probably German – would have done so. Then what? Would the discovery have been kept secret? How soon would weapons development have begun? As it was, physicists understood at once the implications of fission. Fortunately for her, and us, it was Meitner who made the discovery.

Once the *anschluss* happened, Meitner understood that she was no longer safe in Germany. She was then a German citizen. She applied for an exit visa and was turned down. This was the paradox faced by the German Jews who stayed too long. The Germans didn't want them, but they would not let them leave. The solution to this paradox, when it arrived, was final. Meitner was saved because she had colleagues outside Germany who were determined to help her at whatever cost. One of them was the Dutch physicist Dirk Coster. By heroic efforts he raised enough money to support Meitner in Holland for a year. By even more heroic efforts he went to Berlin to try to bring her out of Germany. He had only recently learned of her visa problems. The borders had been sealed with explicit instructions to keep scientists from escaping. Coster chose an obscure border crossing where he was able to exert some influence on the border guards. Meitner spent the night before she left at Hahn's house and then departed with ten marks in her purse and a diamond ring that Hahn had inherited from his mother. Meitner was to sell it if her financial situation became desperate. I have never been able to learn whether or not she sold the ring. The job in Holland lasted for a year, but she was soon exploring other possibilities. She would have liked to go to England, but no position she thought was suitable was available. Again, I think that if she had gone to England she would not have participated in the discovery of fission. As it was, she got a position in Sweden in the Nobel Institute for Experimental Physics in Stockholm. Sweden was a neutral country, so communication with

Hahn was relatively straightforward. He visited her once and even consulted with her as to who should be her successor at his institute – something that Meitner did not appreciate.

By December 1938, Hahn was sending Meitner letters about strange results that he and Strassmann were getting in their uranium experiments. He was even asking for her help and suggesting, somewhat naïvely under the circumstances, that they write a paper together. Hahn described something in his experiment that was totally baffling to him. In the detritus produced after the slow neutrons impinged on uranium, instead of some new radium isotope, which he expected, he found something that was chemically indistinguishable from barium. Why was this so surprising? Barium is a metallic element with an atomic number of 56, whereas uranium, if you recall, has an atomic number of 92. Where did this barium come from? Hahn was totally mystified, as was Meitner initially. On the 21st of December she wrote him a brief note beginning, "Your radium results are very startling. A reaction with slow neutrons that supposedly leads to barium! ... But, in nuclear physics we have experienced so many surprises, that one cannot unconditionally say: it is impossible."[3]

On the 21st of December she left Stockholm to spend the Christmas vacation with some friends in Kungälv on the west coast of Sweden. Her nephew Otto Frisch joined her. He had been born in Vienna in 1904 and had taken his Ph.D. in physics from the University of Vienna in 1927. He then came to Berlin to work, taking a flat near his aunt. They played piano duets together. In 1933 he was dismissed because of the racial laws and spent the next few years as an itinerant physicist, spending a year in London, after which he

[3] Sime, op. cit., p. 235.

went to Bohr's institute in Copenhagen, where he remained until 1939. That Christmas, he came from Copenhagen to Kungälv to see his aunt. It seems the day before Christmas they went for a walk in the woods. As Frisch later recounted it, he was on cross-country skis and she was trotting alongside. That morning he had tried to talk with his aunt about some experiments he wanted to do. She did not want to listen. All she wanted to talk about was Hahn's barium. Frisch tried to distract her by suggesting that Hahn might have made a mistake. She said that Hahn would never make that kind of mistake. He was just too good a chemist. What happened next, at least as Frisch told it later, has always puzzled me. As they were walking, the two of them suddenly understood everything. That much I believe. But Frisch recounts that they sat down on a tree trunk and began to calculate on scraps of paper. I have always been puzzled as to why they had scraps of paper and writing implements with them on their walk. Be that as it may, what were they calculating? To understand this we have to back up a bit.

When working with nuclei with few neutrons and protons one can imagine trying to understand their behavior as individual particles. This requires, of course, quantum mechanics, but one can imagine how one would go about doing it. However, uranium has more than 230 neutrons and protons. It is hopeless to try to treat the particles individually. One focuses instead on their collective behavior. Bohr, and others, had done this prior to 1938 using a model in which the nucleus is taken to resemble a liquid drop. A nucleus is not literally a liquid drop. Liquid drops are visible and have densities of a few grams per cubic centimeter. The sizes of nuclei are conveniently measured in units of length called fermis. To get an idea of how tiny this is, I will write out all the zeros. One fermi is $1/10,000,000,000,000$, which is equal to 10^{-13} centimeter.

Nuclear densities are something like 10^{15} grams per centimeter cubed.

But the collective behavior of the neutrons and protons in the nucleus does in some ways resemble the collective behavior of the molecules in a liquid drop. A liquid drop has a shape. So does a nucleus; it is roughly spherical. The shape of the liquid drop is maintained by surface tension, which acts on the outer molecules. Fundamentally this surface tension is a product of electromagnetic forces. In the liquid-drop model of heavy nuclei there is also surface tension, which acts on the outer neutrons and protons. But here there is an essential difference. There are two forces involved. There is a repulsive electric force due to the fact that the protons all have the same positive charge. But then there is an attractive strong nuclear force that keeps the nucleus together. For the heavy nuclei, with their many protons, the two forces are poised on a kind of knife edge. In fact these nuclei split spontaneously at varying rates, something that, as we shall see, played a very important role in the design of weapons. But that is not what Meitner and Frisch were considering. They argued that the incident neutron could trigger this instability and that the nucleus would then split. In February 1939, they published a very brief account of their work in Nature.[4] They wrote, "On account of their close packing and strong energy exchange, the particles in a heavy nucleus would be expected to move in a collective way which has some resemblance to the movement of a liquid drop. If the movement is made sufficiently violent by adding energy, such a drop may divide itself into smaller drops." They gave the name "fission" – a name that Frisch had gotten from a biological colleague

[4] Disintegration of Uranium by Neutrons; A New Type of Nuclear Reaction, by Lise Meitner and Otto Frisch, Nature, 143, 239–240 (Feb. 11, 1939).

in Copenhagen who was interested in cell division – to this process. This was the first and last time, as far as I know, in which the word "fission" was used in the literature in quotation marks. But what were they calculating while sitting on the tree trunk?

The important question was whether this reaction was energetically possible. This came down to whether the masses of the final products of the fission were less than the mass of a uranium nucleus plus the incident neutron. To determine the mass of the uranium nucleus they used the mass of the most common isotope in natural uranium; the kind that comes from a mine. This was uranium-238, which has 92 protons and 146 neutrons. The mass of the uranium nucleus is not quite the sum of the neutron and proton masses because there is an energy loss when these particles are bound together. This is the same sort of thing we saw when we were discussing the bound electron-proton model of the neutron. Meitner had had a research associate named Carl Friedrich von Weizsäcker. We will hear much more of him when we discuss the German atomic bomb. His father was Baron Ernst von Weizsäcker, who was a very high official in the foreign office. When Meitner was trying to leave Germany he was contacted for help, but he was too busy helping to export Jews from Germany to Poland. He was later tried and convicted of crimes against humanity at Nuremburg. He served time in prison. Young Weizsäcker was a very good theoretical physicist. He devised a formula from which the masses of the various nuclei could be computed. This formula made use of the liquid-drop model and took some of the parameters from experiments on known nuclei. Using these parameters one could readily find the mass of any nucleus – at least approximately – if the nucleus was reasonably heavy. The liquid-drop model does not work well for light nuclei in which there are too few neutrons and protons.

That fixed the mass of uranium-238, but what about the nuclei that were produced in the fission? The initial nucleus was uranium, which has a nuclear charge of 92. One of the final nuclei was barium, with a charge of 56. The other one, to conserve charge, must have a charge of 92 minus 56, which equals 36. This is the element krypton. Thus they could use Weizsäcker's formula, which Meitner knew by heart, to compute its mass, add it to the mass of the barium nucleus, and then compare the result with the mass of the uranium and the initial neutron. If you multiply this mass difference by the square of the speed of light, c^2, then you will have whatever excess energy is available after the fission. When they did this they found that there was an excess energy of about 200 million electron volts. We must now ask what this unit means. Is it big or small? Compared to what? Understanding this is crucial in what follows.

The electron volt came into prominence in physics with Bohr's atom. Let us take the case of hydrogen. It consists of a proton nucleus with an electron moving in an orbit around it. In the Rutherford model, this electron could move in any orbit, so it was difficult to see what kept the atom stable. Why didn't the electron simply crash into the proton? But in the Bohr atom only certain orbits are allowed. We call them "Bohr orbits." The orbit of least energy is known as the ground state. Once the electron is in it, it can't go anywhere unless energy is supplied from the outside. We can do this by applying radiation. The question is how much energy is needed to knock the electron from the atom. The minimum energy needed turns out to be 13.6 electron volts. Quanta of visible light do not quite have enough energy to do this. They have between 1.5 and 3.5 electron volts. The electron volt is the natural unit for atoms and the natural unit for chemistry. When, after the discovery of the neutron, the modern era of nuclear physics began, it was realized that the natural unit here was millions of electron volts. To take a specific example, if you bind a neutron and proton together you make the nucleus of "heavy hydrogen" – this will also play an important role in what follows. This nucleus is called the "deuteron." Its binding energy is about 2.2 million electron volts. Now, a million of something sounds like a lot, but what does it amount to in the sort of units we use to run our households? The energy unit here is what is called a "joule." Unless you have been exposed to a physics course this will also be unfamiliar. You will have heard of the "watt." The watt is not a measure of energy but a rate at which energy is used. One watt corresponds to a rate of energy use of one joule per second. Your hundred-watt bulb is dispensing one hundred joules per second of energy. Now we are getting somewhere, and we can ask how many joules there are in a million electron volts. The answer may surprise you. One million electron volts is equal to $1.6/10,000,000,000,000$ or 1.6×10^{-13} joule. You may well wonder how this absurdly small number can possibly lead to any useful nuclear energy, let alone a bomb. I begin to explain that in the next chapter. But here I want to emphasize that an individual nuclear fission supplies a totally negligible energy on any practical scale. Before I get to bombs I want to finish this chapter by telling you what happened immediately after Frisch returned to Copenhagen with the news.

The first thing that Frisch did was carry out a successful experiment that showed that fission produced the pulses that Fermi's group failed to see. Bohr, on the other hand, was leaving almost immediately for the United States. He left by boat on January 7 and arrived in New York on January 16. He was headed to Princeton, where he was going to spend the spring semester at the Institute for Advanced Study. He was accompanied by his assistant, the Belgian physicist Léon Rosenfeld. On the sea voyage Bohr had arranged for a blackboard in his cabin on which he went over the fission cal-culations again and again. He had promised Frisch not to say any-thing before the paper with Meitner was published. But, of course, he told Rosenfeld – but did not tell him to keep it quiet. Bohr stayed a bit in New York while Rosenfeld went to Princeton and told everyone. The news spread like wildfire and within days the fis-sion experiments had been confirmed at various laboratories across the United States and soon in Europe. When the Meitner-Frisch paper appeared in February it was almost anticlimactic. The order of names on the paper is just as I wrote – first Meitner and then Frisch. The blood of scientific priority is thicker than the water of kinship.

4. Chain Reactions

Figure 6. Einstein (left) and Leo Szilard (right). March of Time/Time & Life Pictures/Getty Images.

I T IS IMPORTANT FOR US TO UNDERSTAND WHAT MEITNER AND
Frisch did – and did not – do. This was not understood by the
various Nobel Prize committees who, a few months after Hiroshima,
awarded the prize in chemistry for 1945 to Hahn alone for the discovery of fission. They did not understand what Frisch and Meitner
had done. What they did was to make the first qualitative proposal
as to how fission might occur. What they did not do was to make
a quantitative theory. This was the work of Niels Bohr and John
Wheeler, who was a professor at Princeton when Bohr was visiting the Institute. Their monumental paper – which some members of the Nobel committee incorrectly thought anticipated Frisch
and Meitner rather than vice versa – was submitted to the *Physical
Review* in June 1939.[1] But the previous February, Bohr had published
a brief letter to the editor of the *Physical Review* entitled "Resonance
in Uranium and Thorium Disintegrations and the Phenomenon of
Nuclear Fission."[2] The origins of this letter are interesting. When
Bohr was visiting the Institute there was another physicist there by

[1] The Mechanism of Nuclear Fission, by Niels Bohr and John Archibald Wheeler,
Phys. Rev., **66**, 426–450 (1939).
[2] *Phys. Rev.*, **55**, 418–419 (1939).

the name of George Placzek. Placzek, who was Czechoslovakian, was noted for his skeptic wit. He apparently listened to Bohr's account of the liquid-drop model for nuclear fission and told Bohr that it was all nonsense. Placzek's complaint had to do with the energetics. The claim was that a slow neutron could come into a uranium nucleus and create so much havoc that the nucleus would split. Where did the energy come from to cause this excitation? After hearing Placzek's question, the excitation was Bohr's. John Wheeler, who was present, told me about Bohr's reaction. He was so disturbed by Placzek's question that he headed off into the town of Princeton, pacing rapidly in random directions as Wheeler tried to keep up. At some point in their darting about, Bohr had an epiphany, which, as it happened, changed the entire course of the development of nuclear weapons, although that was the last thing on Bohr's mind. To the contrary, he thought that it made nuclear weapons a practical impossibility. Let me quote the key phrase in Bohr's *Physical Review* letter and then go on to explain it and hint at the consequences, which we shall later discuss in detail. Here is what he wrote: "We have the possibility of attributing the effect concerned [the efficacy of slow neutrons causing fission] to a fission of the excited nucleus of mass 236 formed by the impact of the neutrons on the rare isotope of mass 235."[3]

To understand this we must say a bit more about nuclear structure. The force that holds the neutrons and protons together is of exceedingly short range. To clarify what I mean, the gravitational force is very long range. Although the Sun is ninety-three million miles away, its gravitational attraction keeps the Earth in its orbit. In contrast, the nuclear force has a range of the order of fermis, that is, 10^{-13} centimeter. This means that neutrons and protons

---

[3] Bohr and Wheeler, op. cit., p. 419.

like to interact only with close neighbors in the nucleus. But what if there is the odd proton or neutron with no immediate neighbor? We would expect the neutrons and protons in such a nucleus to be bound together less tightly than in a nucleus in which all the neutrons and protons are paired up. Let us consider two examples. The most common isotope of helium has a nucleus of two protons and two neutrons. This makes the pairing ideal. The average binding energy per nuclear particle – nucleon – is 7.1 million electron volts. As these things go, this is a very high energy. That is why this helium nucleus – the alpha particle – is so stable. It is emitted bodily in some radioactive decays. On the other hand, lithium-7 has three protons and four neutrons in its nucleus. There is an odd neutron out. In this case the average binding energy is just 5.6 million electron volts per nucleon. Helium-3, which again has the odd neutron out, has an average binding energy per nucleon of only 2.1 million electron volts.

But what has this to do with fission? Bohr realized that what matters in the liquid-drop model of fission is not the binding energy of the nucleus before it absorbs a neutron, but rather the binding energy of the compound nucleus after it absorbs a neutron. In the case of uranium-235 the compound nucleus is uranium-236 and in the case of uranium-238 the compound nucleus is uranium-239. In the first case all the neutrons and protons have partners whereas in the second case there is the odd neutron out. Thus there should be more excess energy given off when uranium-236 is formed than when uranium-239 is formed. It is this excess energy that causes the liquid drop to break up. So Bohr speculated that, in natural uranium, all the fission takes place in the relatively rare unranium-235 nuclei.

When Bohr made this speculation the masses of uranium-236 and uranium-239 were not known. They are now. The binding energy per nucleon for uranium-239 in millions of electron volts is 7.559

whereas for uranium-236 it is 7.586 million electron volts, confirming that this nucleus is more tightly bound by what seems like a tiny difference. But this difference makes all the difference. In the formation of the uranium-236 compound nucleus, so much energy is available for the excitation of the nuclear liquid drop that the neutron that instigates the excitation does not need to have any energy at all. In particular, neutrons whose speeds have been moderated so that they are at "thermal" speeds of a few kilometers per second can initiate the fission of uranium-235. On the other hand, it requires at least an initial neutron energy of one million electron volts, corresponding to a neutron speed of about a few hundredths the speed of light, to initiate the fission of uranium-238. An isotope like uranium-235 is called "fissile," whereas an isotope like uranium-238 is called "fissionable." Bohr concluded that essentially all the fissions generated in the Hahn-Strassmann experiment came from the fissioning of the "rare" isotope – uranium-235. How rare is it and why is it rare?

In the Earth's crust uranium is not especially rare. Its abundance is comparable to other metals such as tin and tungsten. There are places in rocks and shales where the abundance is higher. When you mine uranium in such places you will find that, by weight, the percentage of uranium-235 is less than 1 percent – 0.72 percent, to be precise. If a scientist did not tell you, you probably would never guess where this terrestrial uranium comes from. It comes from the stars – supernova explosions that took place some six billion years ago. A complex set of processes takes place in these explosions that involves the absorption of neutrons, building up the heavy elements in the periodic table. It requires fewer neutrons to make a uranium-235 nucleus than a uranium-238 nucleus, so the initial production of uranium-235 was about 1.65 times greater than the production of uranium-238. Then why is it so rare now?

There are nineteen different isotopes of uranium, all of them unstable in the sense that they decay into less massive nuclei. Three of the isotopes are what I would call "meta-stable." They live a very long time. In the radioactivity business the usual measurement of this kind of duration is called the "half-life." This is the time it takes for half of any sample to decay away. In decreasing order of their half-lives, you have uranium-238, with a half-life of 4.5 billion years. Then you have uranium-235, with a half-life of 700 million years, and finally uranium-234, with a half-life of only 250,000 years. Uranium-239, for example, has a half-life of about 23 minutes. The only meta-stable isotope of uranium that is fissile is uranium-235.

Bohr came to the conclusion that because of its rarity this made any practical prospect of nuclear weapons impossible. His reason was that you would need to separate uranium-235 from uranium-238 – something that came to be called "enrichment" because the isotopic content was being enriched. But isotopes are chemically identical, which rules out chemical methods of separation. You would have to make use of the tiny mass difference between the two isotopes, which is essentially the mass of three neutrons.[4] But each neutron has a mass of about 1.7×10^{-24} gram. Hence the mass difference between the isotopes is minuscule. In December 1939, Bohr gave a lecture in which he said, "With present technical means it is...impossible to purify the rare uranium isotope to realize the chain reaction."[5] We turn next to what is meant in this context by a "chain reaction," but this is a view that Bohr held until the summer of 1943. It was then that he came to believe, erroneously as it turned

---

[4] To be more precise, whether it is the actual mass difference or the ratio of this difference to the total mass depends on what method of separation is being used. For centrifuges, for example, it is the absolute mass difference. In any event, for isotope separation these differences, absolute or relative, are minuscule.

[5] See *Niels Bohr's Times*, by Abraham Pais, Oxford University Press, New York, 1991, p. 462.

out, that the German physicist Werner Heisenberg was actually building a bomb. Bohr escaped from German-occupied Denmark that fall, first to Sweden, then to England, then to the United States. He was briefed about the Allies' bomb project in England and then he came to Los Alamos, where he saw what he had predicted to be impossible well on the way to being realized.

In the last chapter I left you with a conundrum. Each fission produces an amount of energy that is minuscule by any practical measure. How then is it possible to use nuclear energy derived from fission to do anything, let alone make a bomb? That is what I am going to discuss now. It will also give me the opportunity to introduce the rather odd figure of Leo Szilard. Szilard was born on February 11, 1898, into a comfortable middle-class Jewish family in Budapest.[6] Their name had been Spitz until his father changed it to Szilard to make it sound more Hungarian. Szilard was a precocious child who spoke and read German and French by the time he was six and who studied English a little later. He was tutored at home until he was ten. He then went to the Lutheran Gymnasium, which was also the school that the physicist Eugene Wigner and the mathematician John von Neumann attended. The three of them would play essential roles in the building of the bomb. After high school Szilard matriculated at the Palatine Joseph Technical University with the intention of becoming a civil engineer. He liked physics but in Hungary at the time about the only thing a physicist could do professionally was teach in a high school. Szilard had a brief stint in the army during the war and then attempted to continue his education at the Technical University. There was an outburst of anti-Semitism

---

[6] For a biography of Szilard see *Genius in the Shadows*, by William Lanouette, University of Chicago Press, Chicago, 1992; and Szilard's papers in *The Collected Works of Leo Szilard*, MIT Press, Cambridge, Mass., 1972.

in Hungary so he decided to emigrate to Germany to continue his education. He enrolled in the Technische Hochschule – the Technical Institute – in Berlin, still with the intention of becoming an engineer. However he was finding the engineering curriculum increasingly dull. Physics was being taught at the Friedrich-Wilhelm University in the center of Berlin. One day he wandered into the physics colloquium.

Szilard may not have realized that at the university and the associated institutes there was assembled what was probably the greatest collection of physicists ever assembled in one place. Los Alamos in its heyday was a close candidate for second. To name a few, Max Planck, who first introduced quanta, was there, as was Einstein. Less known, but also of great importance, was Max von Laue. Von Laue, who was born in 1879 – making him the same age as Einstein – had won the Nobel Prize in Physics for 1914 for his work on crystals. Von Laue has always been one of my heroes. He was the only physicist I know of who remained in Germany and who took an active stance against the Nazis. Why he was not arrested and sent to a concentration camp I cannot imagine. Von Laue was one of the people who helped plan Meitner's escape. After the war, a physicist who was going to Germany asked Einstein if there were any German physicists he wanted to be remembered to. Einstein said, "Give my greetings to Laue." When he was asked if there was anyone else he repeated, "Give my greetings to Laue."

After listening to the colloquia, Szilard decided to change institutions and to study physics. Von Laue agreed to supervise him for his Ph.D. and gave him a problem in relativity to do. Von Laue had written the first technical monograph on the theory. After working on the problem for a month or so Szilard decided he did not like it. He was afraid to tell von Laue, but he came up with his own problem – something in thermodynamics. He wrote a short paper and was again afraid to show it to von Laue, but he decided to show it to Einstein. He first explained to Einstein what the problem was and Einstein said it was impossible. Szilard described his solution and Einstein was impressed and told him to show it to von Laue. Von Laue was also impressed and it became Szilard's thesis in 1922 and his first published paper in 1925. Not long after he completed his thesis Szilard did a second piece of work – the most important piece of pure theoretical physics that he ever did – which he did not make public until 1926. This was the resolution of what was called the paradox of "Maxwell's demon."

In 1867, Maxwell introduced this "being," as he called him, briefly in a letter. Later the "being" became known as a "demon." In 1871, Maxwell adumbrated what he had said in his letter in his monograph on the theory of heat. This so-called demon was a way to violate the second law of thermodynamics, which says that in any process entropy can never decrease. In particular, you cannot transfer heat from a cold to a hot body without doing work. Maxwell proposed a being that could, by looking at any molecule in a gas, tell how fast it was going. Using this information the demon would open a gate only to, say, fast molecules. That way they would all be collected on one side of the partition, raising the temperature of the gas there without apparently having done any work on the system – a violation of the second law. Szilard realized that there was a fallacy in Maxwell's assumption that the information-gathering demon had, when it came to entropy, a free lunch. He argued that the information itself carries entropy and, if this is taken into the balance, the second law is saved. Decades later this was rediscovered and became the basis of what came to be called "information theory." Szilard did not bother to publish his paper, but in 1926, he had to give what was called a "habilitation" lecture. This was a sort of second Ph.D. that was required in many European countries before one could teach in a university. Einstein, for example, had to have one before he could teach at the University of Bern. Three years later Szilard published his lecture and it was more or less forgotten for the next two decades.

At this point Szilard was able to become a *privat dozent* – a sort of private instructor paid directly by the students. More importantly, he became von Laue's assistant. He stayed in a junior position at the university until 1933, becoming friends with people like Meitner and continuing his growing friendship with Einstein. The two of them even took out patents on things like refrigerators. Szilard shared with his contemporary Hungarian colleagues – von Neumann and Wigner, and later Edward Teller – a sort of foreboding about the future in general. Perhaps it came from their experience of being caught up in the political turmoil in Hungary. I have an ineluctable memory of being persuaded by Wigner to take him to meet William Shawn, the editor of the *New Yorker*. This was in the 1970s, and Wigner had become convinced that there was going to be a nuclear Armageddon and that we should all build air-raid shelters. He and the previously mentioned John Wheeler had built them for their families in Princeton. Shawn was not persuaded and, so far at least, Wigner's pessimism has turned out to be wrong. What Szilard was pessimistic about in 1933 in Germany was the future of the Jews. He had, by virtue of his university position, become a German citizen and by then he wanted out. He managed to get to London, where he helped to create an organization called the Academic Assistance Council to help other refugees. On what he was living at this time, or at most other times, is unclear. He mainly lived out of suitcases. It was at this time that he had his epiphany, which is why I have introduced him here.

On September 11, 1933, Rutherford gave a lecture for the British Association for the Advancement of Science on the prospects that the discovery of the neutron had brought about. At the end of his lecture Rutherford speculated on the future of atomic energy and concluded that it had none. He was quoted as saying, "We might in these processes [atomic transformations] obtain very much more energy than the proton supplied [he was talking about accelerating protons], but on the average we could not expect to obtain energy this way. It was a very poor and inefficient way of producing energy, and anyone who looked for a source of power in the transformation of the atoms was talking moonshine."[7] Szilard had a cold and did not attend the lecture, but he read an account of it the next day in the newspapers. For some reason Rutherford's proclamation got on his nerves. A week or so passed and then on a walk in London Szilard had his epiphany.

Suppose, he argued, there was some sort of process initiated by a neutron that produced some energy but, in addition, produced neutrons. Let us say, for the sake of argument, that two neutrons are produced. We see in the next chapter that this is a good choice. This means that in the next generation four neutrons are produced and in the next, eight, and so on. The number grows exponentially. What happens after eighty generations? The number of neutrons has now grown to 2 to the power 80, or 2^{80}. But this is equal to about 10 to the power of 24 – 10^{24}. But that is about equal to the number of nuclei in a kilogram of uranium-235. Now you begin to see the plot. Szilard did not know anything about fission, but let us follow this using what we do know. We argued that each fission would produce roughly 10^{-11} joule of energy. This means that 10^{24} fissions would produce about 10^{13} joules of energy. So we get out about 10^{13} joules per kilogram from the fission of uranium-235. Compare this to TNT, which produces about a million – 10^6 – joules per kilogram. In terms of tons, this is about a billion – 10^9 – joules per ton of TNT. Thus a kilogram of fissioned uranium-235 produces roughly the equivalent of *ten thousand tons* of TNT. (The more detailed calculation found in the Units and Sizes section gives closer to twenty.) Szilard did not know this, of course, but what he did know, or intuit, scared him half to death.

The first thing he did was to file patents. The earliest of these is dated in June 1934. In it he discusses in a rather unspecific way the notion of what he calls a "chain reaction." As far as I know, this is the first time the term was introduced in this context. The specific reaction he proposes using beryllium certainly does not work. But he mentions the possibility of an explosion and even discusses criteria for the amount of material you would need to make a chain reaction self-sustaining – the "critical mass." Whether all this adds up to a viable patent for, say a nuclear weapon, I am not sure. However, Szilard was sufficiently concerned with the implications that he assigned the patent, to be kept secret, to the British Admiralty. He then became a kind of traveling salesman, trying to sell his idea and to warn of possible nuclear explosions. No one was much interested. In 1935, he visited the United States and spent a little time at Columbia University. He managed to get on the nerves of I. I. Rabi, Columbia's leading physicist, by telling everyone the important experiments that they should be doing. Rabi finally told him that if they were all that important he should do them himself. Then, in early 1939, the discovery of fission was announced. That changed everything. There was all of a sudden a mechanism that, as we shall see, produced both energy *and* the neutrons. Szilard, who was by then settled in the United States, was almost in a state of panic. Fission had been discovered in Germany and the consequence must have been, he felt, as evident to the Germans as it was to him. The first thing that Szilard thought to do was to try to persuade his mentor, Einstein, to write to the queen mother of Belgium, who was an old friend of Einstein's. He had known both her and her late husband King Albert I from the time he first attended scientific meetings in Brussels before the First World War. The king had died in a mountaineering accident in 1934, but Einstein still retained his friendship with the then queen mother. Szilard's idea was that by contacting the Belgian government, they might prevent the Germans from exploiting the uranium deposits in the Belgian Congo.

In the summer of 1939, Einstein was spending his vacation in a rented cottage near the water in Peconic on Long Island. Szilard did not know how to drive, but on July 12th he persuaded Wigner to drive him there. The two Hungarians had a bit of a problem asking directions to Peconic, as they were not sure how the Indian name was pronounced. When Szilard explained to Einstein about fission and chain reactions, Einstein said that that was something that he had never thought of. Indeed, Einstein had said that the use of nuclear energy was like hunting at night for birds in a country where there were very few birds. Einstein remarked to Szilard and Wigner that using this kind of energy was the first time that mankind had used energy that did not come directly, or indirectly, from the Sun. They did not realize that it came from the stars.

Einstein dictated a letter to the Belgian Ambassador in German.[8] However, after returning to New York, Szilard discussed the matter

[8] I am grateful to Walter Isaacson for information about these meetings.

with Alexander Sachs, who was a close advisor to President Roosevelt. Sachs said that Einstein should write directly to Roosevelt. Szilard began drafting a letter that ran more than four pages. He mailed it to Einstein, who thought that it needed work. He changed the focus from uranium deposits to a general warning about chain reactions. He suggested a second meeting with Szilard. This time Wigner was not available, so Edward Teller became the chauffer. Szilard had written Einstein that he was sure that Einstein would like to meet Teller because, "He is particularly nice."[9] The two Hungarians again drove to Peconic in Teller's 1935 Plymouth on July 30th. There then occurred several days of corrections and emendations until, on August 2nd, the letter to Roosevelt was written. It began,

Sir,
Some recent work by E. Fermi and L. Szilard, which has been communicated to me in manuscript, leads me to expect that the element uranium may be turned into a new and important source of energy in the immediate future.…

Signed
A. Einstein[10]

[9] Lanouette, op. cit., p. 201.
[10] The full letter can be found in, for example, *Albert Einstein*, by Albrecht Fölsing, Penguin Books, New York, 1997, p. 711.

5. MAUD

Figure 7. Otto Frisch. Wheeler Collection, courtesy AIP Emilio Segré Visual Archives.

T HAT NEUTRONS WOULD BE PRODUCED ALONG WITH FISSION
fragments – krypton and barium in the Hahn-Strassmann
experiment – was not unexpected. If one looks at the table of isotopes, the ones with the greatest stability have the same number of
neutrons and protons in their nuclei. The helium nucleus, with two
neutrons and two protons, is an example that comes to mind. But
the stable carbon nucleus has six neutrons and six protons, and the
stable oxygen nucleus has eight neutrons and eight protons. Heavy
nuclei such as that of uranium-235 cannot achieve this balance.
Uranium-235, as we have mentioned, has 92 protons and 143 neutrons. It is "neutron rich" and also unstable. When uranium is fissioned, some of these excess neutrons are shed. The neutrons that
are produced this way are called "prompt neutrons" because they
appear about 10^{-14} second – instantaneously – after the fission process starts. The isotopes produced in fission are in general unstable,
and their decays can also produce what are known as "delayed neutrons" because they do not appear until something like ten seconds,
or more, after the fission. Whereas delayed neutrons are important
in the design of reactors, they are very much fewer in number than

the prompt neutrons – about 1 percent – and are not important for what I am considering.

Both barium and krypton have innumerable isotopes, and the number of additional prompt neutrons depends on which isotope is produced. To take an example using the notation I introduced earlier for isotopes,

$$^{235}_{92}U + n \rightarrow\ ^{236}_{92}U \rightarrow\ ^{144}_{56}Ba + ^{89}_{36}Kr + 3n.$$

To put this in words, a uranium-235 nucleus absorbs a neutron to become a uranium-236 compound nucleus. It fissions into an isotope of barium and an isotope of krypton. The superscript on the top left tells us the total number of nucleons in the respective nuclei. If we add the initial neutron we have 236 on the left-hand side of the relation. Because neutrons do not disappear, we must end up with 236, which is why in this particular fission there are three prompt neutrons produced. Both barium-144 and krypton-89 are unstable. Not only are these not the *only* fission possibilities, but barium and krypton isotopes are not even the most *likely* fission products. Hahn and Strassmann found barium because that is what they were set up to look for. Figure 8 presents a curve with two humps that shows all the possibilities and their likelihood of occurring. An isotope in the left hump is matched with one of the isotopes in the right hump. You can see that barium, with A = 144, and krypton, with A = 59, are nowhere near the maxima in the humps. I show this only to demonstrate how complex a process fission is. There are at least sixty different radioactive isotopes generated in uranium fission. In these various fission processes, between zero and six, or more, prompt neutrons are produced. Averaging over everything, about 2.5 neutrons are produced. This does not mean that sometimes half a neutron is

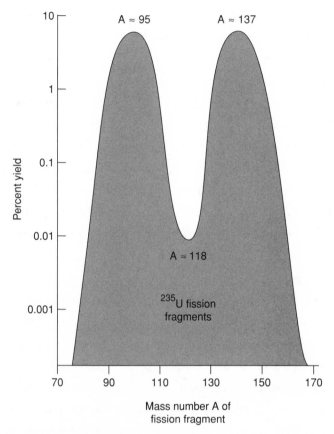

Figure 8. The distribution of fission fragments for uranium-235. In a given fission there is one fragment from each hump.

produced, but 2.5 reflects the average – enough for a chain reaction. Incidentally, most of the energy produced – the 200 million electron volts we discussed earlier – is carried off by the heavy fission fragments. The neutrons carry off rather little.

When fission was discovered, Szilard was living in New York, again out of his suitcase, in the King's Crown Hotel near Columbia. On what he was living is again not clear. He had had a part-time job at Oxford, but that source of income had dried up. Fermi was a professor at Columbia. But Szilard was not in that loop, so he only heard about fission secondhand from Wigner, who had gotten the news in Princeton. Szilard immediately saw the consequences. He was almost desperate to see if the extra neutrons were actually produced, which would provide a physical realization of what he had imagined in his patent. He managed to scrounge some private money to do the experiment and was given space at Columbia. He also got a collaborator – a young Canadian-born physicist named Walter Zinn, who was teaching at the City College of New York. On the 3rd of March they first observed traces of the prompt neutrons. One might take this date in 1939 as the date when an atomic bomb first seemed a real possibility.

Fermi also had the idea of looking for the neutrons and he had enlisted the help of a young Columbia physicist named Herbert Anderson. When Szilard looked in on their experiment he discovered that they weren't finding the neutrons. He had a momentary feeling of relief because he thought he and Zinn might have made a mistake and the neutrons weren't there. But he soon realized the problem. Fermi was using a neutron source that was producing neutrons that were too energetic. He apparently had forgotten the lesson of his own slow neutrons. Szilard gave him a source that produced slower neutrons and the fission neutrons appeared.

The question then was what should be done with this discovery. As far as Fermi was concerned this was just another scientific discovery and should be treated that way, which meant publication in a journal like the *Physical Review*. As far as Szilard was concerned this was a military secret and should be treated that way lest the Germans learn about it. He had persuaded Fermi, when a French group led by Joliot found the neutrons and published despite Szilard's urgings. So Szilard and Zinn published a short letter followed by a long paper, and Anderson, Fermi, and Szilard also published a long paper. This was the paper that Szilard had shown Einstein in manuscript form. Fermi's work also alarmed the chairman of the Columbia Physics Department, George Pegram, who happened to have connections with the navy. He arranged for Fermi to go to Washington to make a presentation. Some idea of the success of this can be gathered by how the desk officer went into the admiral's office to announce Fermi's presence. "There is a wop outside," he was overheard to say.[1] The navy put up $1,500 for further fission research, which was con-siderably less than Szilard had been able to scrounge from private sources. Except by Szilard, the whole matter, at least in the United States, was not taken very seriously. Things were quite different in Germany and Britain. I will come to Germany later, but first, Britain.

The first scientist in Britain to take these matters seriously was not British. It was the aforementioned Otto Frisch. When we last saw Frisch he had just come back to Copenhagen with the news that he and his aunt had understood that the Hahn-Strassmann experiment was the discovery of fission. In the summer of 1939, Frisch went to Birmingham for what he thought was going to be a summer visit. The war broke out and Frisch decided it would be better not return to Copenhagen. It was a wise choice because the Germans occupied Denmark the following April. The head of the physics department in Birmingham was one of Rutherford's for-mer students – an Australian named Mark Oliphant – and offered Frisch a job. Once settled, Frisch decided he wanted to continue his work on uranium fission. In particular, he wanted to test Bohr's

[1] The quote can be found in *The Making of the Atomic Bomb*, by Richard Rhodes, Simon and Schuster, New York, 1986, p. 295.

theoretical argument that it was only uranium-235 that was fissile. To this end he needed to separate the isotopes – to "enrich" the uranium. He chose a method that had been invented by Klaus Clusius in Germany and had been used by Clusius and Gerhard Dickel to separate isotopes in 1938. In its simplest form, the Clusius separation method involved putting a hot wire in the middle of a tube that contained the isotopic mixture in gaseous form. This meant that the outside walls of the tube were substantially cooler than the wire. The theory then predicted that the lighter isotope would collect at the hot wire and the heavier isotope at the cooler surface. Moreover, the lighter isotope should rise to the top of the tube and the heavier isotope collect at the bottom. Clusius and Dickel had demonstrated that this worked for some gaseous mixtures – not uranium. Clusius had also argued that replacing the wire by an inner-heated thick rod would be even better. Frisch asked the laboratory glassblower to make him such a rod. This took more time than expected because the glassblower was busy with requests from people who were working on radar, which had a much higher priority. In the meantime Frisch paid a visit to Liverpool, where James Chadwick was then living, in order to get from Chadwick uranium hexafluoride, which he was going to separate.

A uranium atom can attach itself to six fluorine atoms, producing a compound called "uranium hexafluoride." At room temperature this takes the form of a solid grey crystal that can react violently with water and is corrosive with metals. It is a most unpleasant substance. At higher temperatures it becomes a liquid and then a gas. In its gaseous form it is used in almost all of the uranium separation methods. When one discovers that a country is stockpiling uranium hexafluoride, one can be sure that that country is contemplating uranium separation. In his delightful memoir, *What Little I*

Remember,[2] Frisch describes his encounter with Chadwick. "At the physics department of the university we were shown to Chadwick's study; he came in after a short while, sat down at his desk, and started to scrutinize us, turning his head from side to side like a bird. It was a bit disconcerting, but we waited patiently. After half a minute he suddenly said, 'how much hex do you want?' That was his way; no formalities, straight down to brass tacks."[3] Frisch must have made a good impression because he was invited to do his experiment at Liverpool. Some time later, he found that the Clusius method gave no separation for gaseous uranium hexafluoride. Later in the war the liquid form was used successfully as one of the methods that separated uranium for the Hiroshima bomb.

Before Frisch knew any of this he had a vision. Suppose you could arrange, say, a thousand Clusius tubes in a cascade and thus actually produce pure uranium-235, how much would you need to make a self-sustaining chain reaction, and would it be explosive? The latter question is important. We have seen how much energy would be released if you fissioned a kilogram of uranium. But how long does this take? If it took a millennium, one would hardly consider the event an explosion. If it took fractions of a second, however, that would be another matter. Bohr himself had been stopped by this issue. He had only considered natural uranium, in which neutrons have to be slowed down to enhance the fission reaction with the rare uranium-235 isotope. Thus the whole process would be so slow that, at worst, it would be like setting a fire. But uranium-235 is fissile, meaning neutrons of any speed can fission it. This should decrease the time to fission the kilogram dramatically. But how much would

[2] *What Little I Remember*, by Otto Frisch, Cambridge University Press, New York, 1979.
[3] Frisch, op. cit., p. 132.

Figure 9. Rudolf Peierls. AIP Emilio Segré Visual Archives, Physics Today Collection.

you need – the "critical mass" – to make the chain reaction self-sustaining? Frisch was not a theorist, but he was able to make use of a method that had been devised by a French physicist named Francis Perrin. He did not know experimentally all the parameters to put into the formula but he made plausible guesses. He discovered to his amazement that it would require only a few pounds. This put the whole nuclear weapons issue in an entirely new light, and Frisch immediately shared this information with his Birmingham colleague Rudolf – later Sir Rudolf – Peierls (see Figure 9).

Peierls was born in Berlin in 1907, making him a couple of years younger than Frisch. He had studied at the University of Berlin then in Munich, and then Leipzig, where he got his Ph.D. with Heisenberg in 1928. Then he was off to Zurich to work with Wolfgang Pauli. Pauli once remarked that Peierls spoke so fast that by the time he had understood what he was saying, Peierls was already explaining to him why it was wrong. During this period Peierls helped to create the foundations of the physics of condensed matter – the theory of solids. In 1930, he went to a conference in the Soviet Union and met the physicist Yevgenia Kanegisser – "Genia." They were married in 1931. As I can attest from personal experience, Genia Peierls was a force of nature. I spent the academic year of 1971–72 in Oxford, where Peierls had gone in 1963 to build up the theoretical physics group. I do not think that I had my bags unpacked before Mrs. Peierls was organizing my entire social life.

When Peierls was in Leipzig, Heisenberg had just made the first step in creating quantum mechanics. After the discovery of the neutron, Heisenberg was one of the first people to apply the quantum theory to the nucleus. Peierls had considerable respect, perhaps too much as it turned out, for the ability of the Germans to exploit nuclear energy. In 1932, Peierls won a Rockefeller Foundation fellowship for a year. He spent the first six months in Rome working with Fermi's group and the next six months in Cambridge. During this time the Nazi racial laws were enacted and Peierls realized that he could not go back to Germany. He then found a temporary job in Manchester, where he stayed until 1937. By that time Oliphant was chairman of the department at Birmingham and offered Peierls a professorship there. This is the position he held before he went to Oxford. During this time the "prof," as he was known, had an array of students on the doctoral and postdoctoral level that reads like a

who's who of twentieth-century theoretical physics. John Bell and
Freeman Dyson are two that immediately come to mind.

In 1939, when Frisch was beginning his uranium experiments,
neither he nor Peierls were British citizens. In fact, technically
they were enemy aliens and risked deportation to Canada. Oliphant
seems to have prevented that. But they were not allowed to work on
any classified project such as radar. In the case of Peierls, Oliphant
got around that by posing abstract problems in electromagnetism
that he thought Peierls might have thought about or might want
to think about. Peierls knew that these were radar connected and
Oliphant knew that Peierls knew, but the fiction was maintained.
The work that he and Frisch were about to do on fission had no
classification because there was really no project. That came after
they did their work and then, for a time, they were not allowed to
see the consequences. The two of them began their work in March
1940. They typed up a two-part report themselves, so that no one
else could see it. The first part they called "Memorandum on the
Properties of a Radioactive 'Super-Bomb,'" and the second they
called "On the Construction of a 'Super-Bomb'; Based on a Nuclear
Chain Reaction in Uranium."[4] They were then not sure what to
do with the reports so they showed them to Oliphant. He passed
them on to Henry Tizard, who was a physical chemist and an advisor to Churchill on radar. Tizard passed them on to G. P. Thomson, J.J.'s son, who was chairman of a committee studying the possible utility of chain reactions. They had concluded that there were
none and were in the process of disbanding when the two reports
reached Thomson. The committee was revived with consequences
I will describe after I explain what was in the two reports.

---

[4] The full report with commentary can be found in *The Los Alamos Primer*, by
Robert Serber, University of California Press, Berkeley, 1992, pp. 79–88.

The first report, the one entitled "Memorandum on the Properties of a Radioactive 'Super-Bomb,'" is both graphic and nontechnical. If it had not been followed by the second technical report it might have read like science fiction. It begins,

The attached detailed report [the second memorandum] concerns the possibilities of constructing a "super-bomb" which utilises the energy stored in atomic nuclei as a source of energy. The energy liberated in the explosion of such a super-bomb is about the same as that produced by the explosion of 1,000 tons of dynamite. [This was something of an understatement.] This energy is liberated in a small volume, in which it will, for an instant, produce a temperature comparable to that of the interior of the sun. The blast from such an explosion would destroy life in a wide area. The size of this area is difficult to estimate, but it will probably cover the centre of a big city.[5]

They go on to say that such a weapon would be "practically irresistible,"[6] by which they meant there would be no way to protect oneself against its effects. They argue that the radioactive fallout would kill many civilians and that "this may make it unsuitable as a weapon for use by this country."[7] Ironically, both of them went to Los Alamos where they, indeed, helped build the bombs that did kill many civilians. In the report they express their fears that the Germans could be building such a weapon. They note that Clusius might have been building an industrial facility to use his process to separate uranium. They did not then know that his method cannot separate gaseous uranium hexafluoride. I would imagine that the British weapons scientists must have received innumerable proposals for "death-rays" and the like – weapons that would end the war. What made this proposal different was the second memorandum,

[5] Serber, op. cit., p. 80.
[6] Serber, op. cit., p. 81.
[7] Serber, op. cit., p. 81.

in which the science is spelled out in such a way that no physicist could fail to understand its implications.

I am going to divide my discussion of this report into two parts. In the first part I will discuss the aspects of the report that did not work out, and in the second part I will discuss the aspects that did. The latter are by far the more important, but we can learn something about nuclear weapons, which is our aim after all, by seeing where Frisch and Peierls missed the mark.

First of all, they spend a good deal of time discussing the Clusius method for separating the isotopes. They even give the dimensions of the tubes and note that to produce a 90 percent enrichment something like one hundred thousand such tubes would be needed. They add helpfully, "This seems like a large number, but it would undoubtedly be possible to design some kind of system which would have the same effective area in a more compact and less expensive form."[8] They clearly do not yet know that this system using gaseous uranium hexafluoride will not work at all. In fact, Peierls went on to invent one that would. In this system the gas is forced through billions of microscopic holes in a membrane. The lighter isotope travels somewhat faster through the holes and is, at least momentarily, concentrated on the other side of the membrane. It can be used as stock for the next stage. This method was also invented in the United States and eventually the two groups collaborated in making it into a useful system.

Frisch and Peierls understood in a general way how an actual bomb might work. Using arguments that I am going to discuss shortly, they had estimated the critical mass of the uranium. In these arguments, and in their design of a weapon, they assumed that the

[8] Serber, op. cit., p. 87.

uranium is in the form of a sphere. They note that if the sphere is initially disassembled into parts that are subcritical, then these parts are not explosive. The idea then is to assemble them very rapidly into a sphere that has at least the critical mass. Their proposed method of doing this reminds me of Leonardo's flying machines, which have the same antique charm. Here is what they say: "A sphere with radius of less than about 3 cm could be made up into two hemispheres, which are pulled together by springs and kept separated by a suitable structure which is removed at the desired moment."[9] When I describe how such subcritical units are assembled in an actual bomb, you will appreciate the delightful simplicity of this picture. It most probably would have produced what is known in the trade as a "fizzle" – a sizable explosion by most standards but not a full-scale nuclear weapon.

Finally, there is the matter of how the chain reaction is initiated once the critical mass has been assembled. To initiate the chain reaction, neutrons have to be injected at just the right time into the critical mass. Frisch and Peierls propose to use the neutrons produced by chance in ambient cosmic rays. This is about as efficient as the springs. Solving the problem of the "initiator" was one of the several difficult technical achievements carried out at Los Alamos. I now want to turn to what they got right, which is much more important and influenced the next stages of the nuclear weapons program in both Britain and the United States. First I want to discuss the critical mass.

In what follows I will, with Frisch and Peierls, assume that the mass is in the form of a sphere. In the bomb that was dropped on Hiroshima that was not the case and the critical masses had to be

[9] Serber, op. cit., p. 86.

determined largely by experiment. What is really relevant in the spherical case is the volume of the critical sphere. If we know the volume, we can find the mass by multiplying the volume by the mass density – the number of grams per centimeter cubed, say. But the volume is proportional to the cube of the radius (r^3), hence it all boils down to finding the critical radius. The first question to ask is why such a radius exists. To understand this we have to consider what happens to one of the prompt neutrons that is produced in a fission. There are essentially three things that can happen to such a neutron when it encounters a uranium-235 nucleus. I will list them in increasing order of probability. The least probable is that the neutron is captured with the emission of radiation. I will ignore this improbable possibility. The next most probable thing is that the neutron can induce fission of the uranium-235 nucleus. The most probable thing is that it can bounce off the nucleus elastically – elastic scattering. It takes roughly five of these collisions before there is a fission. Now we can see the problem. If the neutron scatters out of the sphere before it causes a fission, it is lost to the chain reaction. We can ask, on the average, how far the neutron goes before it causes a fission. This is called the "mean free path" for fission. Now the dimensions of the situation are clear. If the radius of the sphere is smaller than this mean free path, there is a good chance that the neutron will escape before it induces a fission. If it is larger than the mean free path, there will very likely be a fission, but we may have employed more of the difficult-to-produce uranium-235 than we need. Thus the critical radius is just about the size of the fission mean free path. But what determines the fission mean free path?

One obvious determinant is the number of uranium-235 nuclei per centimeter cubed. The bigger this number is, the more nuclei are available to interact with the neutron and the smaller the fission mean free path will be. We can find this number in the following way. The density of uranium is about 18.9 grams per centimeter cubed. This is a high density compared to other elements. For example, the density of iron is about 7.87 grams per centimeter cubed whereas the density of lead is 11.34 grams per centimeter cubed. To get an idea, if we take the Frisch-Peierls hypothetical sphere of uranium with a radius of three centimeters – a tiny sphere – it weighs on the order of a pound! The mass of an individual uranium nucleus is about 4×10^{-22} gram. If we divide the mass density by the mass of one nucleus we will get the number of nuclei in a cubic centimeter, which is about 5×10^{22} nuclei per cubic centimeter.

The second thing that determines the fission mean free path is more subtle. It is the relative probability that the neutron causes a fission as opposed to something else, such as elastic scattering. Physicists measure this probability in terms of an effective area, which they call a "cross section." The bigger this area, the more likely it is, say, for the fission to take place. At this point Frisch and Peierls were stuck. These cross sections were being measured in the United States, but thanks to Szilard and others, the data was kept secret. All that Frisch and Peierls could do was make a guess. They knew that quantum mechanics places an absolute limit on the size of these cross sections, so they chose the largest cross section allowed by the limit. But the bigger this probability, the more likely there is to be a fission, and the smaller will be the mean free path. By taking this maximum cross section they reduced the length of the mean free path. They came up with 2.6 centimeters, whereas the correct answer is about 16.5 centimeters. This meant that the volume of their critical sphere was too small, as was their answer for the critical mass. They found a critical mass of about six hundred grams, whereas the true answer is about 52 kilograms. This is a big difference. One wonders what their attitude would have been if they had come up with the correct answer. It is one thing to have to separate less than two pounds of uranium and quite another if you need to separate more than one hundred pounds.

They also considered whether there would be an explosion rather than just a slow burn. This is a question of the time it takes to fission, say, a kilogram. The whole point of using uranium-235 was that fission is possible with fast neutrons. The neutrons produced in this fission have an average speed of something like a tenth of the speed of light – about 10^9 centimeters a second. How long does it take for one of these neutrons to travel the distance of the mean free path for fission? In doing this I will use the correct answer for the mean free path and not what Frisch and Peierls used. Because this correct distance is longer than what they used, the time between fissions is also longer. We see that it is about 16.5 centimeters divided by 10^9 centimeters per second – that is, about 2×10^{-8} second. This time, 10^{-8} second, plays such an important role in weapons design that the Los Alamos people gave it a name – the "shake" – one shake of a lamb's tail. But we saw that it takes about eighty generations to fission a kilogram. In time this means about eighty shakes. Thus, putting these numbers together, it takes about one microsecond – a *microsecond* – to fission the entire kilogram. You will have an explosion for sure. Another question that Frisch and Peierls examined was why you can't make a chain reaction with natural uranium, which is more than 99 percent uranium-238. The answer that one often reads is that uranium-238 is not fissionable. As I mentioned earlier, this is wrong. It can be fissioned by neutrons that have energies of more than one million electron volts. When a uranium-238 is fissioned, neutrons with a spectrum of energies are produced. Although about three quarters of these have energies higher than the one-million-electron-volt threshold, most are slowed down below it by collisions, so that they cannot cause fissions and produce a self-sustaining chain reaction. Natural uranium cannot be used to make a bomb.

I want to finish this chapter by first giving the arcs of the lives of Frisch and Peierls and then describing what happened to their report. When Frisch returned from Los Alamos after the war, he worked for the British Atomic Energy Research Establishment at Harwell and he held a chair in Cambridge from which he retired in 1972. In 1960, when she was eighty-two, his aunt, Lise Meitner, moved to Cambridge to be close to her nephew and his family. Frisch played the piano for her and they discussed physics. She died in 1968, a few days before her ninetieth birthday. Otto Hahn had died a few months earlier, but her nephew kept the news from her. Frisch died in 1979. Peierls retired from Oxford in 1974, when he reached what was then the mandatory retirement age. For a few years he continued as a professor at the University of Washington in Seattle. He had some security problems because his wife was Russian and he had been a colleague of Klaus Fuchs, who had spied for the Soviet Union at Los Alamos. These were resolved. In 1986, Genia died and Peierls himself died in 1995.

As I mentioned earlier, the Frisch-Peierls reports were sent to Henry Tizard and a committee called the MAUD Committee was formed. It had six people on it, which included Chadwick and Oliphant. They first met in April 1940, and Frisch and Peierls were made consultants. Using Peierls's idea of diffusion separation, some preliminary uranium separation was achieved by the following December. Meanwhile, Peierls had gotten some cross-section data from the United States and raised his estimate to about twelve kilograms for the critical mass. He also had the idea that the efficiency of the bomb could be improved if some sort of tamper was put around the uranium sphere that reflected back some of the escaping neutrons into the sphere. All of this was put into the so-called MAUD report in March 1941. The report begins with a very candid statement, "We should like to emphasize at the beginning of this report that we entered the project with more skepticism than belief, though we felt it was a matter which had to be investigated. As we proceeded we became more and more convinced that release of atomic energy on a large scale is possible, and that conditions can be chosen which would make it a very powerful weapon of war."[10] The report goes on to describe the weapon. In particular it notes that the subcritical components would be assembled by shooting them together with high explosives. Gone are the springs and, indeed, this is how they were assembled for the Hiroshima bomb. It describes what was known about the German effort, all of which gives the report some urgency. In July 1941, the committee was disbanded, but not before sending the report to one Lyman Briggs, who was the director of the so-called Uranium Committee in the United States. Having heard nothing from Briggs, Oliphant decided to visit Briggs in Washington, only to find that Briggs had put the report in his safe and had shown it to no one. Oliphant then visited everyone who might have some influence in getting the project started. It worked, and one can probably date the real beginning of the American program to Oliphant's visit.

The reader may well have wondered where the name "MAUD" came from. When the committee was formed a communication arrived from Bohr that ended with the words "AND TELL MAUD

---

[10] See www.atomicarchive.com/Docs/Begin/MAUD.shtml for the report.

RAY KENT." They assumed that it was some kind of code, which they tried desperately to decipher. It was only when Bohr arrived in Britain in the fall of 1943 that they learned that it was a greeting to a woman named Maud Ray, who had helped with Bohr's children, and who lived in Kent.

6. Eka Osmium

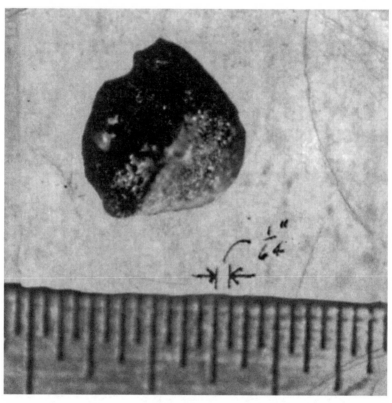

Figure 10. The first gram of plutonium at Los Alamos in March 1944. Courtesy of Los Alamos National Laboratory.

I N 1869 THE RUSSIAN CHEMIST DIMITRI MENDELEEV ORGANIZED
the then-known sixty-three elements into a table in order of
what we would call increasing "atomic weight." For us the atomic
weight is simply a number that is equal to the number of neutrons
and protons in the nucleus. For Mendeleev, who did not know about
atomic nuclei, this represented the weight of the element compared,
say, to hydrogen, which he assigned atomic weight one. Mendeleev
noticed, and he was not the first, that if you arrange the elements this
way they fall into groups that seem to have similar chemical properties. But what immortalized Mendeleev was that he saw that some
elements were missing in the sense that they could be fitted into
holes in his table. This enabled him to predict the atomic weights of
these missing elements and what their properties would be. When,
indeed, some of these elements were found with the properties that
Mendeleev had predicted, his place in the history of chemistry was
assured. One curiosity about Mendeleev's first table is that it does
not have uranium, even though it had been discovered and named

For a more complete treatment see my *Plutonium: A History of the World's Most
Dangerous Element*, Joseph Henry Press, Washington, D.C., 2007.

after the planet Uranus, by the self-educated German chemist Martin Klaproth in 1789. In his later tables Mendeleev included uranium as the heaviest element. Mendeleev was a professor of chemistry at the university in St. Petersburg. He had a colleague named Otto Böhtlingk who taught Sanskrit. Mendeleev learned some of that language – at least how to count. In Sanskrit the words for "one," "two," and "three" are "eka," "dvi," and "tri." Mendeleev used this numbering system to name the missing elements. For example, the element one over from boron in his table he called "eka boron." It was discovered in 1879 by Lars Fredrick Nilson and named "scandium" after Scandinavia. Once the structure of the nucleus was understood and the role of isotopes clarified, the periodic table was organized in terms of increasing nuclear charge and not increasing atomic weight. Figure 11 shows the periodic table as it existed just before the Second World War. It has some interesting features.

Note that the organization is in terms of nuclear charge – the number of protons in the nucleus. Elements in individual columns all have very similar chemical properties. For example, look at the last column to the right beginning with He, which is helium. All the elements in the column such as Ne – neon – and Kr – krypton – are what are often referred to as "noble gases." Their "nobility" consists of the fact that they do not bond readily with other elements. In the modern theory of the periodic table this is explained by the way the electrons are configured around the nuclei of these elements. Notice that there are blanks in the row that has uranium – ^{92}U – blanks for the transuranics that would presumably follow uranium: 93, 94, and so on. Look, for example, at 93. It is just below ^{75}Re, which is rhenium. In much of the scientific literature of this period, element 93 was called "eka rhenium" and its neighbor 94 was called "eka

1																		2
1 H																		2 He
3 Li	4 Be											5 B	6 C	7 N	8 O	9 F	10 Ne
11 Na	12 Mg											13 Al	14 Si	15 P	16 S	17 Cl	18 Ar
19 K	20 Ca	21 Sc	22 Ti	23 V	24 Cr	25 Mn	26 Fe	27 Co	28 Ni	29 Cu	30 Zn	31 Ga	32 Ge	33 As	34 Se	35 Br	36 Kr
37 Rb	38 Sr	39 Y	40 Zr	41 Nb	42 Mo	(43)	44 Ru	45 Rh	46 Pd	47 Ag	48 Cd	49 In	50 Sn	51 Sb	52 Te	53 I	54 Xe
55 Cs	56 Ba	57–71 La–Lu	72 Hi	73 Ta	74 W	75 Re	76 Os	77 Ir	78 Pt	79 Au	80 Hg	81 Ti	82 Pb	83 Bi	84 Po	(85)	86 Rn
(87)	88 Ra	89 Ac	90 Th	91 Pa	92 U	(93)	(94)	(95)	(96)	(97)	(98)	(99)	(100)				

57 La	58 Co	59 Pr	60 Nd	(61)	62 Sm	63 Eu	64 Gd	65 Tb	66 Dy	67 Ho	68 Er	69 Tm	70 Yb	71 Lu

Figure 11. The periodic table before World War II.

95

osmium." The implication was that these elements would behave chemically like rhenium and osmium respectively. This turned out to be wrong. The first people to discover this were two physicists from the University of California at Berkeley, Edwin McMillan and Philip Abelson. I put the names in this order because McMillan was the senior investigator.

There were two reasons, I think, why McMillan and Abelson succeeded in finding the first transuranic where so many others, from Fermi onward, had failed. The first was that fission had been discovered. The second was that they were able to make use of the cyclotron. A lot of the confusion of the early experimenters came about because what they thought represented the radioactivity of the transuranics was really the radioactivity of the fission fragments. Once fission was discovered this was straightened out. The cyclotron was the invention of the Berkeley physicist Ernest Lawrence. He was born in 1901 in South Dakota and had earned his way through St. Olaf College selling pots and pans door to door. He took his Ph.D. from Yale in 1925, and then three years later he decided to accept a job at Berkeley, which was building up its physics department. Robert Oppenheimer soon joined Lawrence there. The idea of accelerating electrically charged particles using high-voltage discharges was familiar to Lawrence. In fact, protons already had been accelerated to an energy of one million electron volts. These accelerators were all "linear." The charged particles moved in a straight line. This limited them because you could only build evacuated tubes of limited length and, to improve on the acceleration, you would have to build one several meters long. Lawrence's idea was to have the charged particles move in circles – or rather spirals of circles with increasing radii. To maintain them in their circular orbits you would apply a magnetic field in the same way that magnetic

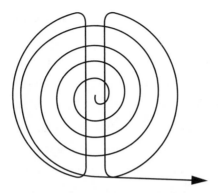

Figure 12. A drawing of the original cyclotron design.

fields were used to separate isotopes in a mass spectrometer. But Lawrence also realized a rather paradoxical sounding fact. The time it takes for such a particle to go around a circular orbit does not depend on the radius of the orbit. The reason for this is that the radius of the orbit depends on the speed of the particle. The faster the particle is moving, the larger the circle it moves in, if you keep the magnetic field fixed. So while a particle moving in a large circle has longer to go, it is going faster in just such a way that the time remains the same. This gave Lawrence the clue as to what to do. If you look at the diagram in Figure 12 you will see two evacuated chambers that look like the letter "d" back to back. In the cyclotron language these are in fact called "dees." Across the gap between the two dees an electric potential is set up. When the particle comes to the gap it is kicked across with this potential and accelerated. But then it comes to the gap a second time and the potential is reversed and kicks the particle across again. The timing of the electric field reversal can be arranged at a fixed frequency, because the time it takes to go around an orbit does not depend on the radius of the circle. This way you avoid the size limitations of the linear accelerator. The limitation in the cyclotron is just the size of the dees, and these can be built larger and larger.

The first cyclotron was built under Lawrence's direction in 1931 by a graduate student named M. Stanley Livingston (yes, this was his real name). The diameter of the dees was about 4.5 inches – almost a toy. But it did accelerate protons to energies of eight thousand electron volts. A short time later a new model was built with a diameter of eleven inches that accelerated protons to one million electron volts. Throughout the Depression, Lawrence was able to raise money to make larger and larger cyclotrons. One of the arguments he used was that he could use the cyclotrons to make isotopes that had great value for medicine. His brother was a doctor. When McMillan began his work in 1939, he had available to him a thirty-seven-inch cyclotron.

McMillan had a specific use in mind for the cyclotron. He wanted to create an intense beam of high-energy neutrons. Fermi and his successors all had to use neutrons that were created in the radioactive decay of unstable nuclei, which limited the numbers and energies of the neutrons that were available. To achieve his objective, McMillan used a process called "deuteron stripping." Remember that the deuteron is the nucleus of heavy hydrogen with one neutron and one proton. If you make a beam of deuterons – which is what McMillan did with the cyclotron – you can use them to bombard a target of, say, beryllium (McMillan's target). Then either the proton or the neutron can be stripped off the deuteron. If the neutron is stripped off, the result is a proton and another isotope of beryllium. But if the proton is stripped, you end up with an isotope of boron and a rapidly moving free neutron, which is what McMillan wanted.

I have, incidentally, a rather affectionate feeling for this kind of process. When I was a graduate student I spent a summer as a sort of intern at the Harvard cyclotron. The experiment I worked on was the inverse of this one. The process was called "deuteron pickup." The Harvard cyclotron accelerated protons, and when these impinged on a target they could pick up a neutron and create an outgoing deuteron. The Harvard cyclotrons had an interesting history. The first one went into operation in 1938. In 1943, it was purchased by the government for one dollar and shipped off lock, stock, and barrel to Los Alamos. After the war a new cyclotron was built with funds from Harvard and the navy. This was the cyclotron I worked on. It was – by comparison to the modern accelerators, which can only be operated by teams of specialists – a rather home-spun affair that the graduate students could run. We did things like stitch together targets and pile lead bricks for shielding. Gone are those days.

When McMillan began his experiments the news about fission had just come to Berkeley. It had created enormous excitement, and McMillan wanted to do a rather simple fission experiment. He knew that after the uranium fissions, fission fragments with a good deal of kinetic energy would be created. He wanted to get a han-dle on this by seeing how far they would go in sheets of aluminum foil that he had stacked together like pages in a book. He then did a second experiment in which he replaced the aluminum foil with stacks of cigarette paper. This should have been pretty mundane except that the cigarette foil experiment revealed an unexpected source of radioactivity with a half-life of about 2.3 days. Along with this there was a second, and expected, source of activity: the beta decay of the uranium-239 isotope that McMillan had created when he bombarded the uranium-238 with neutrons. To McMillan it was perfectly clear what had happened. When the uranium-239 beta decays, a neutron is turned into a proton. The half-life of this process is about twenty-three minutes. When this happens, an isotope of the transuranic-93 – eka rhenium – is produced. He had succeeded in creating the first transuranic element. The problem was proving it. To this end he enlisted the help of a rather recent arrival at Berkeley, the Italian physicist Emilio Segré. He had been part of Fermi's group and, as a Jew, had lost his university job when the Mussolini racial laws were enacted. In 1937, before he came to the United States, he had participated in the discovery of element 43-technetium. This was the first new element, as opposed to an isotope, that had been produced artificially. If you look at the table of elements in Figure 11 you will see that it is the blank element below manganese. In fact it was called "eka manganese." This dates the table. Beneath it in the column you will find rhenium. It has, in general, the same chemistry as technetium. McMillan thought that Segré would be perfect for studying the property of the new element – eka rhenium – which was thought to have the same properties as rhenium. Segré was a little too perfect. He concluded that because the object with the 2.3-day half-life that McMillan had observed did not have the same chemistry as rhenium, it could not possibly be "eka rhenium." It must therefore be another fission fragment. He said so in no uncertain terms in a short note he published.

For a while McMillan accepted this, but then it began to bother him. It did more than bother him when he repeated his experiment using a new sixty-inch cyclotron. He discovered that whatever was producing the 2.3-day activity barely moved – as compared to the fission fragments – after it was produced. It must be something much more massive than a fission fragment, which has a mass of one of the elements somewhere in the middle of the periodic table. McMillan decided that he would do his own chemistry and, in that enterprise, he enlisted a new collaborator, Philip Abelson. Abelson had been a student at Berkeley and, after obtaining his degree, had gone east to work at the Carnegie Institution in Washington. In the spring of 1940, he decided to go back to Berkeley for his vacation. He never thought that he would be spending his vacation finding a new element. As it happened, he had been working at Carnegie trying to separate a small sample of the 2.3-day decay substance from a large amount of uranium. He had not gotten very far with this but had some ideas as to how to do the chemistry. When this was applied to the much more copious samples that McMillan had produced, there was no question that the chemistry of the 2.3-day object resembled that of uranium and had nothing to do with rhenium. This was completely unexpected. They did not know at the time that this element and uranium were members of a new chemical series that came to be called the "actinide" series. This series has the remarkable property that the chemical nature of these elements remains sensibly the same as we move along the row. This is totally different from, say, the noble gases, whose chemistry is the same when we move *down* the column. There was no question that what McMillan and Abelson had discovered was a new element. They published this result in a brief note that appeared in the July 1940 *Physical Review*. I suppose, because of the short half-life of this isotope of element-93 – 2.3 days – they were not worried about military applications. (There is a long-lived isotope that can be used for nuclear weapons.) But there is an odd twist. McMillan had chosen a name for the new element – "neptunium" – because Neptune is the next planet outside Uranus. But they did not use this name in their letter, nor the chemical symbol they had chosen, "Np." It was just "element 93." The name and the symbol were not revealed until August 1945, when the so-called Smyth Report – the account by the Princeton physicist Henry Smyth of the entire nuclear weapons program – was published.[1] Smyth simply notes that this is the name "now used," with no comment.

McMillan understood clearly the next step. The isotope of neptunium that he had manufactured – neptunium-239 – beta decayed with the 2.3-day half-life. This meant that a neutron was turning into a proton, leaving behind an isotope of the element 94. This element was known as "eka osmium" as it was in the column in the periodic table directly below osmium. He conjectured, correctly, that this element would have an alpha-particle decay into an isotope of uranium. He set himself the task of finding this decay. But the signal was too weak. McMillan, again correctly, argued that this meant that this decaying isotope of 94 had a very long half-life. It was a very slow decay. He had some ideas of how to proceed next, but he was recruited to work on radar at the Radiation Laboratory in Cambridge, Massachusetts. Abelson went back to Washington, so the baton was passed to a new player – the Berkeley chemist Glen Seaborg.

Seaborg, who was born in Ishpeming, Michigan, in 1912, was the son of Swedish immigrants. His mother tongue was Swedish. It was not an affectation when in the Nobel Prize ceremony of 1951 – he shared the prize in chemistry with McMillan – he addressed the king in Swedish. In 1922, his mother decided that she had had enough of the winters in Michigan, so the family moved to California, just south of Los Angeles. He attended public school in Watts. His interest in chemistry was aroused by a high school teacher. By this time the family was struggling financially but, as a California resident, he was able to attend the University of California at Los Angeles

[1] *Atomic Energy for Military Purposes*, by Henry Smyth, Princeton University Press, Princeton, N.J., 1948.

tuition-free. He then went to Berkeley for his graduate work and never really left except for government service. After he took his Ph.D. he was able to stay on as a postdoctoral in chemistry, which is when he began working for a brief time with McMillan. They jointly made the unsuccessful attempt to find the alpha-particle decay from the presumed isotope of element 94. After McMillan left for Cambridge, Seaborg decided that he would try to make a different isotope that might have a shorter half-life. Seaborg's idea was to use the deuteron beam directly, as opposed to making neutrons with it. He irradiated uranium-238 with the deuterons from the cyclotron and produced a reaction that gave neptunium-238 plus two neutrons. This was a different isotope from the neptunium-239 that McMillan had produced. This isotope beta decayed with a half-life of 2.1 days into a new isotope of element 94. As he had hoped, it had an alpha decay into uranium with a half-life of 80 years, as opposed to the 24,000 years that McMillan's isotope had. This was in February 1941. There was no question in Seaborg's mind that he had produced element 94. He and his collaborators wrote a letter to the *Physical Review* and even named the element. Because Pluto was the next planet beyond Neptune they called the element "plutonium" and chose the symbol "Pu." But there was a problem – a very serious problem.

Seaborg's long-lived isotope was plutonium-239. It was clear to Seaborg, using Bohr's argument for uranium-235, that this isotope was very likely fissile. In March they did an experiment to measure its fissionability. They discovered that it was even more fissionable than uranium-235. This meant that plutonium was potentially a nuclear explosive. Moreover, separating it from the uranium matrix in which it was produced would require chemical methods that did not depend on the minuscule mass differences between isotopes. In short, plutonium was a very dangerous element. Thus Seaborg sent his letter to the *Physical Review* with the instructions that it not be published until after the war. Throughout the war the name "pluto-nium" was never mentioned. At Los Alamos it was given the code name "49." Its real name was also first revealed in the Smyth report. Until after the war (and, in the case of the Russians, until well after), no one in our program knew that the utility of plutonium in nuclear weapons was known to both the Germans and the Russians. On July 17, 1940, the aforementioned C. F. von Weizsäcker delivered a five-page document to German Army Ordnance, which was then running the official German nuclear energy program. In this docu-ment he proposed using eka rhenium – neptunium – as a nuclear explosive. But by the summer of 1941, Weizsäcker had modified his proposal in a patent application replacing element 93 with element 94. He did not know the names that had been chosen by the dis-coverers of these elements. There is no indication that he, or any of the other Germans, knew anything about the American program. With the Russians it was a different matter. As I will discuss in more detail later, they had an agent in place, Klaus Fuchs, and he told them everything. Throughout the war we were blissfully unaware of any of this.

In the fall of 1941, President Roosevelt was given a copy of a National Academy of Sciences report on the prospects for nuclear weapons. This had been inspired by the MAUD report and work done by our own scientists. On December 7th Pearl Harbor was attacked, and on January 19, 1942, Roosevelt officially authorized the program. It was still a civilian program located in various places from the east to the west coast, even though after 1942 it was actually called the Manhattan Project. The Army Corps of Engineers had its offices in Manhattan. One of the most important physicists running it was a man named Arthur Compton, a University of Chicago professor who had won the Nobel Prize in 1927. There were people who wanted the program on the east coast and people who wanted it on the west coast. Compton split the difference and located it in Chicago. This was decided in January and by April, Seaborg, along with Fermi and Szilard, were in Chicago. Their job was to transform plutonium from a laboratory curiosity to an industrial product. To give an idea of what was going to be involved, all of Seaborg's original experiments had been done with micrograms of plutonium – millionths of a gram. This was possible because the detectors of radioactivity were very sensitive. Although the critical mass of plutonium was not yet known, it would surely be in kilograms. Thus one would need a billion times more plutonium than what was then in hand. The Chicago people did two absolutely essential things. Fermi and his group built the first nuclear reactor, which was a prototype for the plutonium production reactors that followed. Seaborg and his group also found the physical and chemical properties of plutonium, which turned out to be immensely complicated, but which one had to know in order to make use of the stuff. I will begin with Seaborg.

When he and the group he assembled began their work, almost nothing was known about the physical and chemical properties of plutonium. For example, its density was not known. To understand why this was important let me continue an argument I began when I explained how Frisch and Peierls got approximate values for the critical mass of uranium. You will recall that we started by estimating the critical length. I suggested that this varied inversely to the density. In symbols, if I call the critical length r and the density d, then

$$r \sim 1/d.$$

But the critical volume varies as the cube of the radius, so

$$V \sim r^3 \sim 1/d^3.$$

But, to find the mass, we have to multiply V by the density d. Thus M, the critical mass, varies according to

$$M \sim 1/d^2.$$

What this means, to take an example, is if you double the density you reduce the critical mass by a factor of four. This relationship between critical mass and density plays an essential role in the design of nuclear weapons. Now back to Seaborg. The plutonium he was working with in Chicago came from cyclotrons; the one at Berkeley and another at Washington University in St. Louis. This one was kept running twenty-four hours a day, seven days a week, for a year, irradiating uranium to make plutonium. After a year and a half of operating, the two cyclotrons managed to produce two milligrams of plutonium – the size of a grain of salt. But this plutonium was embedded in its uranium matrix. It had to be separated. This involved some difficult and dangerous chemistry. If there had not been a war on, and if they did not think that they were in a race with the Germans, they would not have done this work under the conditions that existed, which were very unsafe. They found that if they dissolved the uranium-plutonium matrix in a witches' brew of sulfuric and nitric acids, laced with a dose of barium phosphate, a crystal was produced that could then be treated further with nitric acid and other ingredients, to produce almost pure plutonium. There was no recipe for this. It had to be done by trial and error using amounts of plutonium that were microscopic. At the end of the process one was left with minute amounts of plutonium metal whose density could in principle be measured.

To measure the density the scientists were able to enlist the services of one of the foremost crystallographers in the world, William "Willie" Zachariasen. Zachariasen, whom I got to know a little, was born in 1906 in Langesund, Norway. His father was a sea captain, and Willie looked like what I thought a Norse sea captain might look like. When he was a boy, he used to row out to the islands near his home and study the crystals found there. A little later in life, he often rowed his teacher at the university, a noted crystallographer named Viktor Goldschmidt, out to a small island that Goldschmidt had bought to preserve its crystals. Willie took his Ph.D. in 1922, the youngest person in Norway up to that time to have done so. Two years later, he was offered a job at the University of Chicago, which he accepted and never left. Willie's specialty was X-ray crystallography, a field that had been pioneered by von Laue. When X-rays are scattered from crystals they produce a pattern that reflects the regularities in the crystal structure. Looking at these patterns, an expert can readily deduce the structure and determine the density of the crystal. One of Willie's specialties was doing this for tiny samples that had been ground up into a powder. He was allotted one hundred micrograms of the precious cyclotron plutonium and asked to find its density. This should have been, for Willie, a piece of cake. But there was trouble. The first time he did the experiment he found a density of thirteen grams per cubic centimeter. When he repeated the experiment he found densities of fifteen and then fifteen and a half grams per cubic centimeter. If someone else had been doing these experiments one might have said that they had made a mistake. But Willie did not make mistakes like this. Something was going on.

Their first idea was that the process of separating plutonium from the uranium matrix had introduced impurities into the plutonium. One thought was that the alpha-particle decay of plutonium might have interacted with these impurities and produced neutrons. These neutrons can cause a pre-detonation of the chain reaction before a critical mass is assembled. Seaborg estimated that you would have to have less than one part in a hundred billion of such impurities to avoid this. This looked like a formidable, perhaps hopeless, problem, but two things happened. In the first place it turned out that impurities were not the cause of Willie's results, rather something more fundamental. In the second place, a much worse problem with using plutonium for bombs presented itself. I explain this in the next chapter, but here I want to explain what Zachariasen learned. Plutonium has a variety of what are known as allotropic forms. Let me give a familiar example of allotropism (if that is a word). Take carbon. There are two allotropic forms of carbon that you are familiar with – graphite and diamond. These have very different physical properties. You can use graphite to make pencils and you can use diamonds to make engagement rings. Which form manifests itself depends on external conditions of temperature and pressure. But they are both forms of carbon. The crystal structures of different allotropes are very different. When it comes to allotropes, plutonium takes the cake. Between room temperature and the melting point of plutonium, which is at 1182.9 degrees Fahrenheit, there are no fewer than six allotropic forms. Each has its own physical properties, including density.

Crystallographers label these allotropic phases with Greek letters; the earliest letters corresponding to the phases of lowest temperature. The phase stable at the lowest temperature is called the alpha phase. For plutonium, the first phase is found at room temperature. It makes a transition to the next, or beta phase, at a temperature of about 234 degrees Fahrenheit. The alpha phase has a density of about 19.8 grams per centimeter cubed. It has a rather unsymmetrical crystalline structure. It is hard and brittle. If you were to try to fold it, it would crack like glass. The next phases are the beta and gamma phases. For nuclear weapons we are interested in the next phase after these – the delta phase. This sets in at about 601 degrees Fahrenheit and lasts until about 844 degrees. It has a symmetrical crystalline structure and is readily malleable. You should be able to make a nice sphere out of it. It has a density of 15.9 grams per centimeter cubed. What must have happened in Zachariasen's experiments was that, unknown to him at the time, his samples must have contained a variety of allotropic phases. He spent much of the war, and the years after, sorting out these phases. The delta phase looked promising for nuclear weapons, but there was a problem: it was not stable. Under relatively small changes in external conditions such as the pressure, it reverted to the alpha phase. One of the challenges that the Los Alamos people faced, and that I also discuss in the next chapter, was how to stabilize this phase. They even learned how to take advantage of the fact that the stabilized delta phase, under the extreme pressures of a nuclear explosion, would revert to the alpha phase. But the alpha phase has a higher density and hence, from what I said before, a smaller critical mass, something that can be used to enhance the explosion.

As I have mentioned, it was clear to everyone at the time that the tiny amounts of plutonium being generated by cyclotrons were not going to be even remotely sufficient for making the amount of plutonium needed for a nuclear weapon. One needed a vastly larger production facility. This was the nuclear reactor. The first idea for designing such a facility, at least in the United States – the

Germans had had the same realization and had begun their own design attempts – came from the fertile imagination of Szilard. As early as 1939, he had begun thinking of a system that would produce large-scale controlled nuclear reactions. He understood that two of the basic components would be the fuel elements and a moderator to slow down the neutrons. The fuel element that Szilard first imagined was powdered uranium with a graphite moderator sort of mixed in with the powder. Although the two men had a somewhat edgy relationship, Szilard started a collaboration with Fermi to construct such an instrument. They had gotten some $6,000 from a rather somnolent Uranium Committee, which had been reluctantly aroused by a second letter from Einstein to Roosevelt. With this, they purchased uranium and graphite in the form of bricks. These graphite bricks had to be stacked – a very messy job. While Fermi and his crew manhandled the bricks, Szilard hired a burly stand-in to do his share. Szilard was free to think. In the course of this he made a discovery about commercial graphite that was crucial. He learned that the manufacturers routinely put boron in the graphite they sold for use in electric arcs. What the National Bureau of Standards had labeled pure graphite contained boron. But boron, even one part in 500,000, soaks up neutrons like a sponge. If present, it would ruin the use of graphite as a moderator.

In this respect, the case of Walther Bothe, the German physicist who attempted during the war to produce a working cyclotron, deserves mention. By 1941, the Germans had decided to investigate the use of graphite as a moderator. As Bothe was the best nuclear experimental physicist in Germany, he was given the job of finding out how neutrons traveled through graphite. He found that they were being absorbed too often to render graphite useful as a moderator. There is some controversy as to whether Bothe understood the reason for this. From what I have been able to learn, it seems that he did, but that the Germans decided that it would be too expensive to purify graphite. Instead they chose to use heavy water as a moderator. This was a fatal choice because the major source of heavy water was in Norway and, in one of the great sagas of the war, this source was sabotaged. The Germans never were able to obtain as much heavy water as they needed. The Germans lacked a Szilard who did not rest until he got some manufacturers to make boron-free graphite.

The type of reactor that Fermi and Szilard first tried to construct was an example of what is known as a "homogeneous" reactor. The uranium and the moderator are all mixed up together. Szilard realized that this reactor had difficulties in its design. The problem was in the moderation of the neutrons. It is essential that while the neutrons are being slowed down they don't get captured by the uranium. This would take them out of the fission business. But, as luck would have it, at certain energies, the chance of capture jumped up. It was as if at these energies the neutrons fell down a well. Szilard had an ingenious idea. Instead of mixing the uranium up uniformly with the moderator, one could isolate it in lumps or rods kept separated. At the surface of one of these lumps the neutrons with the energies that made them vulnerable would be snatched up, leaving the neutrons in the interior with the correct energy to cause fission. This so-called heterogeneous design was the basis of the first reactor that worked. In 1940, Szilard wrote a long paper entitled "Divergent Chain Reactions in Systems Composed of Uranium and Carbon." By "divergent" he meant self-sustaining. He proposed using spherical droplets of uranium rather than powder. He submitted it to the *Physical Review* with instructions that it not be published until after the war. As far as I can tell, it was never published.

In November 1941, the actual construction of a reactor based on these principles began in a squash court under Stagg Field, the football stadium at the University of Chicago. Perhaps suitably, some members of the football team participated in the heavy lifting. The basic idea was to stack uniformly machined graphite blocks in a neatly configured geometric pile. In fact, Fermi gave the name "pile" to the device. The pile began with a circular layer braced with a wooden frame. The next layer, a smaller circle also braced with a frame, was put down. If it had been completed it would have looked like a rough sphere. But as it happened they did not need the full construction to make it work. Alternate layers had holes drilled into the graphite with three-and-a-quarter-inch diameters in which the uranium-oxide pellets were put. In the end they piled 45,000 blocks with 19,000 holes. The photo in Figure 13 shows one of the levels with the drilled holes. They stopped after fifty-seven layers. The uranium pellets were in the form of two-and-a-quarter-inch cylinders of uranium oxide that were dropped into the holes. The chain reaction was to be controlled by inserting cadmium rods that absorbed neutrons but that could be raised or lowered depending on what intensity was wanted. Unlike the subsequent production or power reactors, this one did not have a coolant – something like water that would circulate among the fuel elements and keep them from melting down. The difference was that this reactor, if it worked, would generate almost no power – at first half a watt – and would run for only a very few minutes.

The events of Wednesday, December 2, 1942, have entered into the mythology of twentieth-century physics analogously to the description of how, on that snowy winter day in December 1938, Meitner and Frisch sat down on a tree trunk in the Swedish woods and figured out how fission worked. The difference is that this event

Figure 13. One of the layers of the first nuclear reactor at the University of Chicago. Courtesy of University of Chicago Library.

was witnessed by a great many people who gave similar accounts. At 8:30 in the morning the entire reactor group, and visitors, assembled in the squash court. They went over what they were going to do and at 9:45 Fermi ordered some of the control rods to be withdrawn. The counters that registered the neutron multiplication began to click. Every fifteen minutes or so, Fermi ordered the control rods taken out a foot or so further. Fermi was impassive. Periodically he would announce what the counters were going to do next. This went on until 11:35, when the counters began running almost out of control. An automatic cadmium safety rod slammed down and brought the process to a halt, at which point Fermi announced that he was hungry and wanted to have lunch. Things would resume at two o'clock. Twenty minutes later the safety rod was reset. The extraction of the control rods continued, with the neutron signal steadily rising. At 3:25 Fermi announced that with the next withdrawal the pile would go critical. He was looking for a constant rate of rise, which would happen at criticality. When this happened he knew that the pile had worked as designed. They let it run for twenty-eight minutes at a power output of half a watt. Then they shut it down. Wigner, who observed the test, had concealed behind his back a bottle of Chianti. He gave it to Fermi, who ordered enough paper cups so that everyone got a sip. Compton got on the phone to call James Conant of Harvard, another Manhattan Project mandarin. He said to Conant, "The Italian navigator has landed in the New World." "How were the natives?" Conant asked. "Very friendly," he was told.

I have often thought about this exchange. I wonder if the assembled group had any more idea of what was about to happen than the natives that Columbus encountered. In September 1942, the army engineer who had built the Pentagon – Leslie Groves – was put in charge of the project and promptly gave it its name – Manhattan Project. By November he had selected the DuPont Company to construct the plutonium-producing reactors. By December he had chosen the Hanford site on the Columbia River in Washington State to build the first reactor. This was all done shortly after Fermi's test was completed. By March they had broken ground for the first reactor – the so-called Hanford B reactor. There were some fifty thousand construction workers who built, among many other things, 386 miles of road. The reactor, which Wigner played a major role in designing, was water-cooled. It took water from the Columbia, circulated it, and sent it back to the river, warmer and radioactive. Testing of the B reactor started in July 1944, and plutonium production began in September 1944. The reactor's thermal power production was about 250 megawatts, compared with Fermi's half watt. (The Hanford B reactor did not produce electricity.) The first shipment of plutonium from Hanford to Los Alamos was on February 5, 1945. But prior to all of this, Groves, as a good engineer, had had DuPont build a demonstration facility in Clinton, Tennessee. This reactor went critical in November 1943, and by April 1944 gram samples were arriving at Los Alamos. They were received by two free-spirited metallurgists named Ted Magel and Nick Dallas, who had come to Los Alamos from Seaborg's Chicago group. Magel and Dallas had the job of reducing the first gram sample to a metal. They were supposed to do this publicly on March 24 in front of a group of dignitaries that included General Groves. They decided that this would put too much pressure on them, so they made the reduction, in private, the night before. They left the sample with a note that read, "Here is your button of plutonium. We have gone to Santa Fe for the day."

7. Serber's Primer

Figure 14. Robert Serber (left) and Robert Oppenheimer (right). © The New York Times. Courtesy of Redux Pictures.

T HE HEAVILY REDACTED 1947 FBI FILE ON ROBERT SERBER begins, "His birth date is March, 14, 1909, at Philadelphia, Pennsylvania. Serber is 5'8", weighs 135 pounds, brown hair, blue eyes. Serber's father David was born in Russia and is deceased, his mother Rose was born in the United States and is deceased. Serber lists M. W. Leaf as his father-in-law. . . . Leaf was born in Russia . . ."[1] The report goes on to tell us that in 1930 Serber got a bachelor's degree from Lehigh University in engineering physics and a Ph.D. in physics from the University of Wisconsin in 1934. It was the height of the Depression and jobs were very difficult to find. But Serber was selected for one of the few National Research Council post-doctoral fellowships. He had planned to go to Princeton to work with Wigner. He was on his way east, driving an old Nash with his wife, Charlotte, whom he had married the year before, when they stopped to listen in on some lectures at a University of Michigan summer school in Ann Arbor. Oppenheimer was one of the lectur-ers. After listening to him, Serber decided that he didn't want to go to Princeton after all, but rather Berkeley, where Oppenheimer

---

[1] FBI File 116–7681, obtained through the Freedom of Information Act.

was. The National Research Council agreed, and the couple headed west. I think the history of the atomic bomb might have been quite different if the National Research Council had insisted that Serber go to Princeton.

He remained a postdoctoral student for two years and then Oppenheimer took him on as his assistant. His job was to explain to students what Oppenheimer had told them, which they often did not understand and were too afraid to ask about. Serber and Oppenheimer did a lot of important work in fields that ranged from nuclear physics to the physics of collapsing stars. They became very close friends. Oppenheimer, who was only four years older, had a simple ranch in the Pecos Mountains near Santa Fe. The physicists close to him would decamp there for the weekend to drink whiskey, ride horses, and talk physics. When he first went there, Serber knew nothing about horses, but Oppenheimer put him and his wife on a pair of horses with some whiskey, graham crackers, and food for the horses and sent them off overnight. Serber used to tell a story of one time when they were all riding together at midnight in a lightning storm. They reached a fork and Oppenheimer said that there was a long way and a short way back. He said that they should go the long way because it was "much more beautiful." It was on his solitary rides that Oppenheimer discovered the fairly nearby Los Alamos mesa. When the time came, he knew that it would be the perfect site on which to build a bomb. In 1938, Serber got an offer of a professorship at the University of Illinois, which he was reluctant to accept. He liked being in Berkeley too much. I. I. Rabi, who happened to be in Berkeley for a visit, told Serber that it was so difficult for Jews to get jobs in universities that he had to take it. It later turned out that Oppenheimer had tried to get Serber a job at Berkeley, only to be told that there were already too many Jews.

It was surely because of the bomb that the FBI began paying attention to Serber and his wife, although once their surveillance started it became a retrospective on their lives, and the lives of their families and friends. For example, the report notes, "His associates at the Universities where he was a student and professor were largely comprised of extremely liberal faculty members and students. Although not as active as his wife, he participated in various alleged Communist front organizations in Berkeley, California and Urbana, Illinois." The redacted files further reflect that an informant in Berkeley, California, stated that practically all of Serber's associates while he was living there were known radicals and that, further, his wife frequently held meetings in their apartment for reported Communist groups and alleged Communist front organizations. While Serber and his wife were in Madison, Wisconsin (this was in 1933!), they associated with individuals generally described as "outspoken liberals."[2] This goes on page after page. What has always struck me about this enterprise is that, while the FBI was scrutinizing the list of "outspoken liberals" with whom the Serbers allegedly had contact, they totally missed the three Russian spies who were actually at Los Alamos during the war. I describe all of this in a separate chapter.

In mid-December 1941, just after Pearl Harbor, Serber got a phone call from Oppenheimer, who had been out east and was returning by train to Berkeley. There was something that he wanted to discuss with Serber but could not do it over the phone. He stopped off in Ann Arbor, and the two men took a walk in the countryside, during which Oppenheimer told Serber that he had recently been asked to take charge of the design of a nuclear weapon. He wanted Serber to join him. It was agreed that Serber would come

[2] FBI, op. cit., Part 1, p. 2.

to Berkeley in April. When he did, housing was provided in a room over Oppenheimer's garage. Oppenheimer had assembled a small group of young theorists in Berkeley – largely his former students and postdoctorals. There was also a project at Berkeley directed by Lawrence to separate uranium isotopes by using a kind of mass spectrometer that Lawrence called a "calutron." It later played an essential role in the isotope separation of the uranium that went into the Hiroshima bomb. Plutonium was not yet in the picture. Serber and Oppenheimer had at their disposal some of the papers from the British group, although not the MAUD report. These papers became their starting point for estimating things like the critical mass.

In the summer of 1942, there was an invited conference in Berkeley that included people like Teller and Hans Bethe. Also that summer Teller first proposed a hydrogen bomb and Bethe spent the summer showing that none of Teller's ideas would work. We will come to the hydrogen bomb later, but let me point out that a sine qua non for a hydrogen bomb is a working fission bomb. No country that has not produced a fission bomb can produce a hydrogen bomb. One of the questions that the group considered was whether the explosion of an atomic bomb could ignite the atmosphere and thus destroy life on Earth. They decided that it could not. The physicist Richard Tolman suggested producing a fission bomb by wrapping the sphere with high explosives and using them to compress the sphere, producing a critical mass by increasing the density. At the time, this idea – implosion – was not followed through but, as we shall see later, it became the essential key for making a nuclear weapon with plutonium. In March 1943, Oppenheimer left Berkeley for Los Alamos. The Serbers followed a few days later. The race to make the atomic bomb was on.

One of the first things that Oppenheimer asked Serber to do was to give a series of lectures – in the event there were five – that would serve as an introduction to the physics of the bomb for new arrivals. It is worth commenting on this. First of all, from a fundamental point of view there was no new physics required to make a bomb. The Los Alamos physicists often said that when they came back from the war they found the same set of problems on their desks that they had left. However, building nuclear weapons required using what was then standard physics in novel ways. This is what Serber explained in his lectures. The lectures were attended by some fifty people and were taken down by the well-known physicist E. U. Condon. Afterward, he and Serber wrote up the notes. They were of course classified. They became declassified in 1965 and were published in book form with Serber's commentaries in 1992.[3]

No doubt because of the secrecy, physicists who had not been involved with the bomb knew very little about it. And even the physicists who did know weren't talking. When I was an undergraduate at Harvard in the late 1940s many of my teachers had been involved. John Van Vleck, who had been Serber's Ph.D. advisor at Wisconsin and then had gone to Harvard, was one of the people who attended the 1942 conference that Oppenheimer organized in Berkeley. Kenneth Bainbridge, who was for a while chairman of the department, had chosen the test site for the first atomic bomb explosion in Almogordo, New Mexico. Norman Ramsey, from whom I took a course, had supervised the armoring of the Hiroshima bomb on Tinian Island in the South Pacific. Roy Glauber, who was just starting his academic career – he shared the Nobel Prize in Physics

---

[3] *The Los Alamos Primer*, by Robert Serber, University of California Press, Berkeley, 1992.

for 2005 – went to Los Alamos immediately after finishing the course work for his bachelor's degree. And the president of the university, James Conant, had been one of the original civilian leaders of the program. Despite all of this, I do not remember a single part of any course in which the physics of nuclear weapons was discussed. I do not even remember any of these people mentioning any of this in any context. I can understand this at the time. The war had been over for only a very few years and the Cold War was just beginning. However, not long ago I decided to make a little survey. I went to a physics library and looked at every textbook of nuclear physics I could find. There was not a single one of them that explained how to find the critical mass of a uranium sphere. Most of them did not even explain what a critical mass was. This disturbed me, especially in view of what seemed to be happening in the world. Anyway, on to Serber's primer.

The first thing that struck me on reading the primer was the opening paragraph under the heading "Object." It states, "The object of the project is to produce a *practical military weapon* in the form of a bomb in which the energy is released by a fast neutron chain reaction in one or more of the materials known to show nuclear fission."[4]

The striking thing about this, it seems to me, is its lack of ambiguity. It makes it clear that the people at Los Alamos were not there to do basic research. They were there to make a bomb.

The first lecture deals with material that we have discussed – how long it takes to fission a kilogram of uranium-235 and what energy is produced. But there is an additional point that Serber makes that is worth noting. In the microsecond it takes to fission a kilogram of uranium, the explosion produces a temperature in the material of about

---

[4] Serber, op. cit., p. 3.

ten billion degrees centigrade (eighteen billion degrees Fahrenheit) if the reaction proceeds at a 10 percent efficiency. The interior of the Sun has a temperature of about fifteen million degrees centigrade (twenty-seven million degrees Fahrenheit). Thus, loosely speaking, the bomb has a temperature of a thousand Suns. The idea of efficiency is very important here. At these temperatures uranium is no longer a metal. It is a very rapidly expanding gas. The particles expand at speeds that approach 10^8 centimeters a second – about one hundredth of the speed of light. As the particles of the gas – the uranium nuclei, for example – get farther from one another, the fission neutrons have a harder and harder time finding other uranium nuclei to fission. In short order they do not find enough, and the chain reaction stops. This happens in a time on the order of hundredths of a microsecond. After this, the remaining unfissioned uranium departs into the atmosphere. The efficiency is a measure of the ratio of how much of the material is actually fissioned relative to the amount that was there to be fissioned. This ratio depends on the design of the bomb. The efficiency of the Hiroshima bomb – "Little Boy" – was somewhat more than 1 percent, whereas the Nagasaki plutonium bomb – "Fat Man" – had about a 17 percent efficiency. Modern designs do considerably better. It is ironic to think that, after the millions and millions of dollars that were spent to separate the uranium isotopes, of the 60 kilograms of, on the average, over 80 percent enriched uranium used in Little Boy, only about 0.7 kilogram of the uranium-235 actually fissioned. The rest – 59 kilograms – went off into thin air.

Serber devotes several pages of his notes to a calculation of the critical mass of a uranium sphere. This is the calculation that I looked for in vain in all those nuclear physics textbooks. As I mentioned earlier, what one actually calculates is the critical radius, from which one can find the critical volume. From the volume, knowing the density of uranium, you can calculate the critical mass. But here I want to explain something about what a critical mass means and what it does not mean. In the popular literature there is the impression that if you could assemble a critical mass of a fissile element you would have Armageddon – a nuclear explosion. But this is not the case. Let us recall how the notion of a critical radius was arrived at. Once the chain reaction is initiated in, say, a sphere of uranium-235, there are two effects. In the interior of the sphere neutrons are being produced. But at the surface some of these neutrons are escaping and are lost to the chain reaction. If the radius of the sphere is smaller than the average length the neutron has to travel between fissions then more neutrons escape through the surface than are created in the interior. If we increase the radius, the balance shifts toward the interior neutron number. The critical radius is just the radius where these two effects balance – as many neutrons are created in the interior as escape through the surface. But this suggests that at this radius there is no explosion. There is an equilibrium. This is correct. In fact, at Los Alamos, experiments were designed to start with some mass of fissile material and then add additional material to see exactly what the critical mass would be. During, and shortly after, the war, this was done by hand and was very dangerous. There were two fatal accidents when, in one way or another, too much mass was added at the wrong time and there was a burst of radiation. Even at the critical mass there was no nuclear explosion. After these accidents the work was done remotely. One of the configurations tested was a bare uranium sphere. That experiment was called "Lady Godiva." In short, to make a bomb, assembling a critical mass of material is necessary, but it is not sufficient. You must assemble a "supercritical" mass.

It is very difficult to calculate the efficiency of a nuclear weapon from anything like first principles. In my view, one of the most difficult-to-understand sections of Serber's primer is his presentation of a much simplified version of these calculations. I will not try to reproduce them here. Rather I want to convey the general idea. In a fission weapon there is a sequence of events. You begin with some unassembled fissile material, each of whose component parts has less than a critical mass. You do not want to handle material at the critical mass for fear of irradiation. You then combine this material as rapidly as possible to make a supercritical mass. This will produce an explosion and the material will expand and become less dense. After a short time you will again reach the critical mass appropriate to the new density. This is called "second criticality." At this point, whereas fission may continue for a time, the chain reaction is no longer "divergent," that is, explosive. It is intuitively clear that the more critical mass you can assemble, the longer it will take to reach this stage. Hence more material will fission and the efficiency will be greater. However, the connection between the amount of supercritical mass and the efficiency is not trivial, even with all the approximations that Serber makes. He concludes that if the supercritical mass is double the critical mass of the separated pieces, you only find an efficiency of about 1 percent. I found, using his formulae, that if you triple the supercritical mass you might get up to about 4 percent. But this raises the question of how you could possibly triple the supercritical mass.

To understand the problem let us recall the somewhat naïve "bomb" that Frisch and Peierls proposed in their first memorandum. They envisioned two uranium hemispheres held together by springs but kept separated. Each of these hemispheres had somewhat less than the critical mass. Now you bring them together, and you

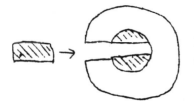

Figure 15. Serber's drawing of a gun-assembly device.

have a supercritical mass of somewhat less than two critical masses. Naïvely, one might think that this is about the best you can do. But never underestimate the ingenuity of physicists. I am going to show you how you can, in principle, produce a configuration with three critical masses. To this end I am going to conduct what Einstein used to call a *gedanken* experiment – a "thought" experiment. I will illustrate it with a diagram taken from the primer (see Figure 15).[5] As the diagram shows, Serber was not exactly a draughtsman.

In my thought experiment I am going to suppose I deal with a material like uranium-235, which at normal densities, d, has a critical mass of M. In my *gedanken* experiment I am going to imagine that I have somehow assembled three critical masses in a sphere. You will object that this will blow up. In my thought experiment I am not going to worry about that for the moment. Now what I do is to remove one critical mass as shown in Serber's diagram. In the diagram the stuff with the diagonals represents the uranium. The white stuff I will discuss later, so, for the moment, pretend it is not there. Thus, in what remains of the sphere, I have 2M worth of uranium. Now comes the trick. What is the density of the sphere with the hole? Well, where there is still uranium, it is d, but where there is nothing it is zero. Thus the average density is $2/3d$ because only two

[5] Serber, op. cit., p. 56.

thirds of what remains of the sphere has uranium in it. But remember what I explained before about critical masses and densities. The critical mass varies as the reciprocal of the square of the density. The example I gave is that, if you cut the density by a factor of two, then the critical mass increases by a factor of four. Here I have cut the density by a factor of two thirds, so the critical mass increases by a factor of nine fourths, which is 2.25. Note, however, that 2.25 is greater than 2. Hence the mass 2M, which is what is left in the partial sphere, is subcritical in this configuration. So, in principle, you would construct such a truncated sphere, which would be nicely subcritical, and then fire the missing piece into it, making a supercritical mass of 3M. This is what Serber's diagram shows. It is the essence of what is known as a "gun-assembly" fission bomb, which is what the Hiroshima bomb, Little Boy, was. The devil was, as they say, in the details.

The gun assembly, with the expectation of using plutonium, was the priority project at Los Alamos in the spring of 1943. Oppenheimer put together a strong team that included the aforementioned McMillan, who had been recruited from his radar work. McMillan's name brings up something – Nobel Prize winners. Every time I have written about Los Alamos I have compiled a list of Nobel Prize winners, some who, like Fermi, Chadwick, and Bohr, had the prize before they went there and others who, like Richard Feynman, got one later. I always manage to leave a few out, or the list gets added to – Glauber being the most recent addition. But what strikes me are the people like Feynman and Ramsey, who were in their twenties, and who later got the prize for things that were not even being imagined by the generation of physicists working at Los Alamos. Never has there been assembled in one place, for one project, a group with

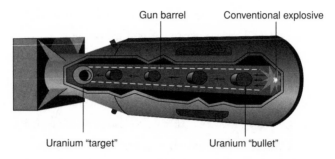

Gun barrel   Conventional explosive

Uranium "target"   Uranium "bullet"

Figure 16. Diagram of a gun-assembly device taken from Serber's primer. This diagram is of course very impressionistic, and in the remainder of this chapter I fill in some of the details. Some of these apply only to Little Boy, but many of them bring up issues relevant to bomb design more generally.

this scientific brilliance. One wonders if it will ever happen again, and for what purpose.

It became clear that the gun assembly involved problems of military ordnance – gun powder, cannon size, and the rest. It was somewhat ironic that the most sophisticated weapon ever designed needed this kind of technology. Oppenheimer realized that the physicists were a little over their heads, so he brought in a naval captain and expert in the field, William Parsons, to head this part of the activity. Three-inch naval guns and twenty-millimeter anti-aircraft guns were made available for testing. Serber floated an idea of two projectiles shooting at each other, but this was deemed too complicated and was replaced by one projectile shooting at a target. The weapon was supposed to look something like what is shown in Figure 16.

The uranium "bullet" in Little Boy consisted of a cylindrical stack of six uranium rings that was about ten centimeters wide and sixteen centimeters long. This stack contained only the most highly enriched uranium – about twenty-six kilograms of about 89 percent uranium-235. This had come from the enrichment facilities in Oak

Ridge, Tennessee, where the first stages used the diffusion methods and the final and largest portion of the enrichment was done using Lawrence's calutrons. This uranium was not delivered to Los Alamos until the beginning of June 1945, and two weeks later the first bullet components were done. The bullet was to be fired into a hollow-cylinder target sixteen centimeters long and containing about thirty-eight kilograms of 50 percent enriched uranium. I must emphasize that these numbers are approximate, as the exact dimensions of Little Boy appear to be classified. The numbers I am using are derived from various reliable open sources and are meant to be orders of magnitude. I do not have access to the classified numbers.[6] It is interesting that in 1948 Fuchs turned over the design of this weapon to the Russians, who, as far as I know, have not declassified it.

The most significant thing I want to emphasize is the times. It took the bullet about half a millisecond to traverse the length of the cylinder and to produce a supercritical mass. A millisecond may seem like the blink of an eye, but in this business it is an eon. Recall that it takes only a microsecond to fission a kilogram of pure uranium-235.

The reason that this time is so important has to do with the phenomenon of spontaneous fission. Heavy nuclei like uranium and plutonium are sort of poised on the edge. As we have seen, they can be made to fission by bombarding them with neutrons, but they can also fission without any external agitation. These fissions produce neutrons. In a nuclear weapon the last thing one wants are stray neutrons wandering around. These can pre-detonate the fissile material by starting a chain reaction before a supercritical mass has

---

[6] See, for example, Wikipedia's article on Little Boy (http://en.wikipedia.org/wiki/ Little_Boy) and the references cited therein.

been assembled. The term of art for this in the weapons business is a "fizzle." Fizzles can be very nasty explosions that spew a lot of radioactive material around, but they are not atomic bombs in the usual sense. Here are a couple of neutron numbers produced in spontaneous fission to give you an idea of how this works. Spontaneous fission of uranium-235 produces something like 0.01 neutron per second per kilogram, assuming that each fission produces something like 2.5 neutrons on the average. On the other hand U-238 produces something like 168 neutrons per kilogram per second. In the mixture of uranium-235 and -238 that was used in Little Boy, only a few neutrons per second were produced. Given that it takes half a millisecond for the bullet to traverse the target, these numbers are sufficiently small so as not to pose a problem for Little Boy. But when we come to plutonium, its numbers will turn out to be a disaster.[7]

In the primer, Serber discusses the role of a "tamper." This term is a little misleading because what is known as a "tamper" serves two roles. In general it is a heavy metal that is put around the fissile material. That is what the whitish parts of Serber's little sketch represent. This metal does two things. It retards the expansion of the exploding uranium and it reflects neutrons back into the fissioning material. Both of these functions improve the efficiency of the bomb but not to the same degree as adding more supercritical mass. The ideal tamper is uranium-238, but it could not be used for Little Boy. It requires so much material to be effective – at least two hundred kilograms – that it would generate about thirty thousand neutrons a

---

[7] The numbers for spontaneous fission have changed a good deal over the years. I am grateful to Cameron Reed and Carey Sublette for discussions of the most recent numbers.

second in spontaneous fission. This ruled out a uranium-238 tamper for Little Boy. Instead, a 2,300-kilogram mixture of tungsten carbide and steel was used. This lowered the critical mass of the unit by about 15 percent. When I discuss the plutonium bombs in the next chapter, I point out that uranium-238 tampers can lower the critical masses there by a factor of two or more. The critical assembly is so fast that the spontaneous fission issue becomes moot.

Another thing that Serber discusses in the primer is the matter of "initiation." How do you start the chain reaction? In Little Boy this could have been done somewhat haphazardly by letting the ambient neutrons, which are always around, start things. But this is not very efficient. Serber suggested making an initiator out of polonium and beryllium.[8] The idea was that polonium – the element that had been discovered and named by Madame Curie – is a copious emitter of alpha particles. When these impinge on the beryllium, neutrons are produced. This was one of the ways experimenters got neutron beams before the cyclotron was invented. In the initiator the two elements are kept shielded from each other until the explosive blast from the bullet arrives. Then a burst of neutrons is emitted and that starts the chain reaction. Little Boy was fitted with a somewhat primitive version. The whole bomb was about 126 inches long, 28 inches in diameter, and weighed about 8,900 pounds.

On July 24, 1945, the last component of Little Boy, the target, was completed. Everything but the target arrived by ship at Tinian Island, in the Marianas, on July 26 and, two days later, the target arrived by air. By July 31, the bomb had been assembled. It could

---

[8] The particular isotope used was polonium-210. It became notorious when it was used to poison the ex–Russian agent Alexander Litvinenko in London in 2007. To manufacture this element in quantity you need to use a reactor.

have been used at any time after that. It was probably the least safe atomic bomb ever made. It had never been tested. If, at 2:45 A.M. on August 6, when the *Enola Gay* took off headed for Japan, the bomber had crashed on the runway, the first test of Little Boy might have been the destruction of Tinian rather than of Hiroshima.

8. The "Gadget"

Figure 17. The "gadget" being lifted to the tower at Alamogordo. Courtesy of Los Alamos National Laboratory.

O N AUGUST 9, 1945, THE PLUTONIUM BOMB THAT I DESCRIBE in this chapter was dropped on Nagasaki. Serber was supposed to have flown on that mission to take photographs, but someone had forgotten to load his parachute, so he had to get off the plane. As soon as the peace treaty was signed with Japan, Serber was sent on a mission to assess the damage to the two cities. He was accompanied by a medical team and a British physicist named William Penney. The two had gone to graduate school together in Wisconsin and later Penney came to Los Alamos as part of the British delegation. After the war, he played a major role in the British atomic bomb program. In describing this experience, which haunted him for the rest of his life, one of Serber's recollections was the fact that he and Penney felt completely safe. There was no sense of menace from the survivors – just polite curiosity. Serber brought home a souvenir. It was a piece of wood that had dark lines in it. It turned out that these were the shadows made when the first light from the bomb entered the windows of a house. The casings on the windows made the shadow. From the angle of the shadows Serber was able to determine the height at which the bomb exploded. It was approximately two thousand feet. An instrument sensitive to air pressure had set it off at the altitude that had been previously decided on. This height had been chosen to maximize the damage caused on the ground – the blast wave. Serber and Penney had brought Geiger counters to measure residual radiation. There wasn't any. Because they had been air blasts, the radioactive debris from the bombs fell elsewhere.

After the war, Serber got a position at Berkeley. In fact, he headed the theory division at Lawrence's Radiation Laboratory. Oppenheimer had left to become the director of the Institute for Advanced Study in Princeton. Perhaps that left room in the Jewish quota, or perhaps the war had persuaded the physics department to change its position. I do not think that Serber did any more consulting on nuclear weapons for the government. In 1948, his clearance was challenged but, despite the 1947 FBI report in which he was labeled a security risk, he was cleared, but not until 1952. In 1949, the Board of Regents of the University of California devised a loyalty oath for the employees of the university. One had to affirm one's loyalty to the state of California and that one did not belong to any organization that advocated the overthrow of the United States government. Some forty employees refused to sign as a matter of principle. These included several very distinguished members of the Berkeley physics department. Serber signed, but the atmosphere at the laboratory became less and less tolerable, so, in 1951, he left Berkeley for Columbia, which is where I met him and Charlotte. Charlotte, who had run the library at Los Alamos, died in 1967. Serber remarried and, after he retired from Columbia, he spent a good deal of time sailing in the Caribbean. He died on the first of June 1997 at the age of eighty-eight. The last time I spoke to him he was in the hospital and he told me that a lifetime of cigarette smoking had finally caught up with him.

Serber's primer does not have a great deal to say about plutonium – "49," as they called it at Los Alamos. This is not surprising. In
the spring of 1943, when the lectures were given, not that much was
known about plutonium. As I mentioned previously, it was not until
a year later that gram quantities began to arrive at Los Alamos from
the Clinton reactors. By the middle of April 1944, Segré's group was
reporting the makings of a disaster. To understand this let me show
the sequence of reactions that lead from uranium-238 to plutonium-
239:

$$^{238}_{92}U + n \rightarrow {}^{239}_{92}U \frac{\beta^-}{23.5 \text{ min.}} \rightarrow {}^{239}_{93}Np$$

$$\rightarrow \frac{\beta^-}{2.33 \text{ days}} \rightarrow {}^{239}_{94}Pu.$$

In words, uranium-238 absorbs a neutron to become uranium-
239, which then beta decays into neptunium-239, which in turn
decays into the fissile plutonium isotope, plutonium-239. So far, so
good. In a cyclotron the neutron flux is relatively low. But a reactor is a neutron factory. There are so many neutrons that if the
plutonium-239 is not rapidly removed it can absorb another neutron and become plutonium-240. In fact about one in three neutrons
that are absorbed by plutonium-239 leads to plutonium-240. The
reactor samples that arrived at Los Alamos were contaminated with
plutonium-240. The reason this was a disaster, as Segré discovered,
was that the spontaneous fission of plutonium-240 produced about
one million neutrons a second per kilogram. Contrast this with
plutonium-239, in which the spontaneous fission produces about
twenty neutrons a second per kilogram. This result ruled out reactor
plutonium as the explosive element for a gun-assembly weapon. The
assembly is just too slow. During the millisecond it takes to assemble

the stuff, so many neutrons are produced by spontaneous fission that a fizzle is all but certain. Serber discusses the problem of spontaneous fission in the primer. But he makes the assumption that the only plutonium that one had to be concerned with was plutonium-239. He conjectured correctly that plutonium-239 would produce a manageable number of neutrons in its spontaneous fission. But neither he, nor anyone else, anticipated the plutonium-240 disaster.

It is said that, upon learning of this, Oppenheimer considered resigning as director.[1] It is not clear what purpose that would have served and, in the event, he did not. Some thought was given to trying to separate the isotopes. But this would have been even more difficult than separating the uranium isotopes. Uranium-238 and uranium-235 differ in mass by roughly three neutron masses, but plutonium-239 and plutonium-240 differ by only one neutron mass. Finally it was decided that the only route to go was "implosion" – the use of high explosives to compress, say, a sphere of fissile material. As I have mentioned, in their 1942 Berkeley meeting, the Caltech physicist Richard Tolman suggested something that vaguely looks like implosion. In the primer, Serber includes a somewhat impressionistic drawing and a description (see Figure 18). The idea, he says in his lecture, was to mount subcritical material on a ring. On the outside would be the explosives, distributed in a sphere. The explosives were to blow the pieces inward to make a sphere. This, as I will explain, is not exactly how an actual implosion device works, but in theory even this version would have the advantage of not having the long time lag while the "bullet" enters the target. When Los Alamos was created, a small amount of work was done on implosion, largely

---

[1] For an excellent description of this and other matters connected with the construction of the bomb, see *Critical Assembly*, by Lillian Hoddeson et al., Cambridge University Press, New York, 1993.

Figure 18. Serber's drawing of an implosion device.

by the physicist Seth Neddermeyer. In 1943, some of the mandarins, such as Conant, asked Oppenheimer what was being done about implosion, and Oppenheimer invented the fiction that Serber was "looking into it,"[2] which was in fact about the last thing that Serber was doing.

The news about plutonium-240 was presented to everyone in a meeting of all the technical personnel at the laboratory that took place on July 4, 1944. By the end of July the entire laboratory, with the exception of the relatively small group that was still working on the gun-assembly uranium bomb, had been mobilized to work on implosion. In order to convey what that meant, here is a no doubt imperfect analogy. Suppose General Motors received the news that all automobiles but hybrids were defective. Could they turn their production around in a few weeks? All sorts of economic issues would come into play. What would the stockholders say? In the case of the bomb there was one "stockholder" – General Groves. He decided that the laboratory would devote itself to implosion, which is what it did. I am going to describe some of the problems the laboratory had to solve, but first I want to convey the general idea in terms of a kind of ideal implosion bomb.

[2] *The Los Alamos Primer*, by Robert Serber, University of California Press, Berkeley, 1992, p. 59.

Let us start with a solid sphere of plutonium metal. I am ignoring all the complications, which I will get into later, of actually making such a sphere. Let us suppose it weighs ten kilograms. This is roughly the critical mass of untampered plutonium. Again I am ignoring all the issues involving the allotropic phases. Let us suppose that 10 percent of this is plutonium-240. This overstates the amount of plutonium-240 in the actual bomb, which was kept to less than 1 percent by extracting the plutonium from the Hanford reactors very quickly. But I want to make a point. Let us wrap a layer of high explosives around the sphere and let us suppose that we can detonate the layer at all points simultaneously. This is clearly an idealization. Let us suppose that the pressure wave from the high explosives can, in something like a microsecond, compress the original sphere into one half of its previous volume. This, in fact, happens in the actual bomb. Now remember that the density of anything is the mass per unit volume. Thus, if we reduce the volume by a factor of two, we increase the density by the same factor. Now we can invoke the critical mass mantra that I have used several times. Namely, if you increase the density by a factor of two, you decrease the critical mass by a factor of four. But see what that means. We started with a sphere that was at about the critical mass for plutonium at normal densities, and in a microsecond, without adding an iota of new material, we have created a supercritical mass. This is the essential idea behind implosion. Note how it solves the plutonium-240 problem. Even if we take the 10 percent figure, which would give us a kilogram of plutonium-240, and we acknowledge that one million neutrons per second per kilogram are produced in spontaneous fission, the fact that the compression takes place in a microsecond makes this neutron production irrelevant. Now I want to turn to the real question: how can you actually make such a bomb?

When Serber began his lectures, he started referring to the object under construction as a "bomb." Oppenheimer got wind of this and said that because there were workmen around it had to be given a code name. "Gadget" was chosen. The first issue I want to discuss about the gadget is the metallic sphere. Enough was known about plutonium in the spring of 1943, when Los Alamos was created, for Oppenheimer to realize that a metallurgy group was essential. Somehow the name of Cyril Smith came to his attention. Smith was born in Birmingham, England, in 1903 and took a degree in metallurgy from the university in 1924, and then a Ph.D. from MIT in 1926. A year later he began working for the American Brass Company in the Naugatuck Valley in Connecticut, where he remained until he went to Washington at the beginning of the war to join what was called the War Metallurgy Committee – a desk job that Smith thoroughly disliked. In February 1943, Oppenheimer had a little chat with Smith on a park bench in Washington. It is not clear what Oppenheimer could have told Smith, who did not have clearance, but a month later, Smith was in Los Alamos creating a metallurgy group. As it turned out, no better choice would have been possible.

A decision was made not to do serious metallurgic work on plutonium until gram samples arrived at the laboratory, which happened in March 1944. It was at this point that Magel and Dallas came from Seaborg's group in Chicago, where they had already developed techniques for reducing plutonium to a metal. Until they appeared, the people at Los Alamos were having a great deal of difficulty doing this. Once Magel and Dallas produced their "button" – which incidentally was the first sample of metallic plutonium that was big enough to actually see without the aid of a microscope – the other metallurgists could start measuring its properties, including the density. They soon confirmed what Zachariasen had discovered about the allotropic phases of plutonium – that there were several and that they had different densities and other physical properties including critical masses. A bare sphere of the room-temperature brittle alpha phase has a density of 19.86 grams per cubic centimeter and a critical mass of about ten kilograms, whereas the higher-temperature malleable delta phase has a density of 15.6 grams per cubic centimeter and a bare critical mass of about sixteen kilograms. Because of the speed of assembly of the implosion weapon, uranium-238 can be used as a tamper. With a tamper several inches thick the critical masses can be reduced by a factor of two.

The delta phase could not be used directly because with the slightest provocation it morphs into the alpha phase. Enter Smith. Smith had spent his entire professional life working with brass, which is mainly an alloy of copper and zinc in various proportions depending on the use. It was very natural for him to think in terms of plutonium alloys. There was no theory to guide him, and as far as I can tell, there still isn't. It was a matter of trial and error. By early spring of 1945, Smith had discovered that aluminum alloyed with delta-phase plutonium appeared to be stable. But there was a catch. Plutonium decays, producing alpha particles and these, impinging on aluminum, would produce neutrons, so you would be back in the trouble that you thought you had escaped from. Then Smith discovered that gallium alloyed with delta-phase plutonium has the same desirable characteristics, without the neutron problem. But, by this time, the test of the gadget was scheduled for only a few months away. There was no time to study for how long the gallium would actually maintain the stability of the delta-phase plutonium. Smith's years of working with alloys gave him a feeling that it would be stable, but the decision to use it was one that only Oppenheimer could make. Oppenheimer decided to treat Smith to a dinner at Edith Warner's little restaurant, which was located in what had originally been the post office at the Otowi railway station. Oppenheimer had come to know Edith Warner before the war. They both had originally come to New Mexico for their health. Before the physicists took over the mesa Edith Warner had served chocolate cake to the boys who came down from the Los Alamos Ranch School. Now she served dinners mainly to people that Oppenheimer invited. Over dinner the two men discussed gallium. Oppenheimer told Smith to make any decision that he thought was right. Smith chose gallium. The gadget consisted of delta-phase plutonium stabilized by 0.8 percent gallium by weight. I should point out that if this alloy is subjected to moderately strong pressure it reverts irreversibly back to the alpha phase. This phase has a higher density and a smaller critical mass. This poses both a risk and an opportunity. The risk is that such a pressure is produced accidentally and hence what one assumed to be a subcritical mass suddenly becomes critical. The opportunity arises because the pressures produced in the atomic explosion cause this reversion to the alpha phase, thereby making the plutonium even more supercritical.

It is one thing to tout the advantages of being able to compress a sphere of plutonium to half its volume in a microsecond, and it is quite another actually to do it. In the summer of 1944, groups at the laboratory began experiments using hollow spheres and cylinders of various heavy metals. There was no thought of using plutonium because there wasn't enough and, besides, if the experiments worked it would have blown up the laboratory. They used shaped charges – charges that are designed to produce their explosive force in select directions – to try to compress the test shapes. The initial results were awful. There was no compression, and jets of explosive energy mangled the objects. In late summer, Robert Christy, who had been a graduate student of Oppenheimer's at Berkeley and whom Oppenheimer had brought into the program even before Los Alamos, had a suggestion. He had done some calculations that seemed to show that the implosion would go much more smoothly if the hollow spheres were replaced by solid spheres. This suggestion was adopted, and henceforth the device was known as the "Christy gadget."

But the real breakthrough came from an unlikely source – the mathematical genius John von Neumann. In the context of Los Alamos one hesitates to use the word "genius." By any standards Fermi was one. So was Oppenheimer. So was Feynman. So was Hans Bethe, who ran the theory division, and on and on. But I think von Neumann was something special – another class of genius. He was born in Budapest in 1903, making him a contemporary of Wigner and Szilard. Like them, he received his elementary education at the Lutheran Gymnasium. Genius in both mathematics and music seems to declare itself early. There seems to have been no stage in von Neumann's life when he was not doing mathematics. He published his first paper at age seventeen. After graduation he was sent by his father to Berlin to study to become a chemical engineer. This lasted for two years, and then von Neumann went to the Eidgenössische Technische Hochschule (ETH) in Zurich, Einstein's alma mater, to complete his engineering education. In the meantime he was getting a Ph.D. in mathematics from Budapest University. By the time he was twenty-two he had two doctorates – one in engineering from the ETH and one in mathematics from Budapest. In 1926, he went to Göttingen, which was then considered the mathematical capital of the world. During the two years he was at Göttingen, von Neumann did something unexpected. He mathematically formalized the newly developed quantum theory. The physicists were speaking in mathematics, without having made

the mathematics of what they were doing very systematic. He sum-
marized this work in his classic *Matematische Grundlagen der Quan-
tenmechanik*, which was published in 1932. The next year he joined
Einstein as one of the first six permanent members of the Institute
for Advanced Study. During the prewar years he worked in almost
every branch of mathematics. His ability to assimilate new mathematics almost instantly was legendary. My teacher in mathematics
at Harvard, the late George Mackey, who also worked on the mathematical foundations of the quantum theory, spent some time at the
Institute. He found a generalization of one of von Neumann's the-
orems. He thought that he might have a pleasant chat with von
Neumann about it. He stated the result and von Neumann asked
for the proof. After two sentences von Neumann said, "I see," and
that was that.

During the war, von Neumann consulted for various military
projects, and I will shortly describe his implosion contribution,
but first I want to mention one other. At the Aberdeen Proving
Ground in Maryland about one hundred women were employed
using mechanical calculators to produce artillery firing tables. This
was being supervised by a young army lieutenant named Herman
Goldstine. He received a suggestion from a professor of electrical
engineering at the University of Pennsylvania named John Mauchly
to try to automate these calculations electronically. They received
a grant and, by 1945, they managed to build the first electronic
computer, the ENIAC. However, in the summer of 1944, Goldstine had a chance encounter with von Neumann on a railway platform in Aberdeen. They began talking and it turned out that the
two of them had similar problems – how to do extremely complicated numerical calculations. Goldstine did not learn until later that
von Neumann's calculations were being done for the Los Alamos

implosion project. Von Neumann became very interested in the electronic computer that was being built, and he joined their study group. In the course of this he formalized the architecture of the modern computer, work that was published in a 101-page memo. Whereas all of the ideas were certainly not those of von Neumann, due to his ability to synthesize everything, the design of modern computers became known as the "von Neumann architecture." It was because of computers that I had my one and only contact with von Neumann. He had come to Harvard to give a series of lectures on the similarities and differences between the computer and the brain. My teacher George Mackey urged me to go. It was the greatest set of lectures I have ever heard. It was like champagne for the mind. Von Neumann spoke without notes, and one lucid sentence followed another. After the first of the lectures I found myself walking in a kind of trance in Harvard Square. When I looked up, there was von Neumann. I thought, correctly as it turned out, that this would be my only chance to ask him a question – he died in 1957, not many years after. In any event, I asked, "Professor von Neumann, will the computer ever replace the human mathematician?" He studied me carefully and then answered, "Sonny, don't worry about it."

The implosion idea, which was first suggested by the British physicist M. J. Poole and then brought to Los Alamos by another British physicist, James Tuck, was so simple that a high school student could have thought of it. To begin with an analogy, imagine a water wave impinging on a sand bar at an angle to it. The leading edge of the wave will hit the shallow water first and will be slowed. The effect of this is that the wave will turn until it is parallel to the sand bar. This is the phenomenon of "refraction." It operates with light. The speed of light in the air is faster than that in glass. So when a light wave passes into a piece of glass, it is bent. If you arrange the glass as

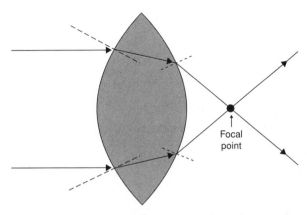

Figure 19. Refraction by a converging lens. Incident rays that travel parallel to the principal axis will refract through the lens and converge to a point.

shown in Figure 19, you can focus the light at a point. This is how you design a lens.

The scientists at Los Alamos realized that what was important here was the different wave speeds in the two media. Different explosives produce pressure waves that move with different speeds. If you conjoined a fast-moving explosive next to a slow-moving one, you would produce a similar refraction effect. Thus the explosive lens was born. Figure 20 provides a schematic drawing of how it

Figure 20. The converging of waves in an explosive lens; the dark wave is the slower explosive. The effect is to make the explosive wave converge so as to compress the plutonium sphere.

works. The spherical explosive wave coming from the right meets a slower explosive and is focused at a point to the left. The British physicists had studied a crude two-dimensional model lens. Von Neumann designed an actual three-dimensional operational lens. Typically he synthesized the work of others into something that was essentially new.

To make something practical out of this idea Oppenheimer needed an explosives expert. He got one of the best, George Kistiakowski. Kistiakowski was born in 1900 in Kiev, Russia. During the Revolution he fought on the side of the "Whites," and then he left Russia to study in Berlin. Perhaps it was his experience as a soldier that gave him a taste for explosives. Even after 1930, when he became a professor at Harvard, he consulted on explosives for the government. In 1942, he was appointed the head of the explosives division of the National Defense Resources Committee in Washington. When Los Alamos was reorganized in 1944, Oppenheimer made Kistiakowski the head of a newly formed X-division that dealt with conventional high explosives. The problems were formidable. It took thousands of experimental castings of the explosives before they were able to make them, with precision, in the desired shape. The eventual design for the Christy gadget had twenty hexagonal and twelve pentagonal lenses. It looked something like Figure 21, which resembles a Bucky-ball, or a soccer ball.

The explosives were mixtures of TNT and other more exotic ones. The pieces had to fit together with a tolerance of no more than a millimeter. Another serious problem was how to detonate the ensemble. This was done by an array of thirty-two detonators that could be fired with a simultaneity to within a few nanoseconds. At the center of this lens array was the plutonium "pit," which was a nearly solid sphere with a radius of about 4.5 centimeters. The

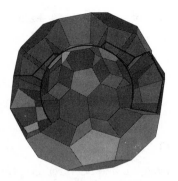

Figure 21. A plutonium pit surrounded by
explosive lenses.

"nearly" had to do with the initiator. In the case of the gun-assembly
weapon, this was almost optional. There were enough ambient neutrons around to start the chain reaction in the slowly assembling
mass. But here, the assembly was being done in a microsecond,
so there was no guarantee of suitable ambient initiating neutrons.
After much work they came up with a much more sophisticated version of Serber's original idea of using polonium and beryllium kept
separated until uniting them by the shock wave from the implosion, releasing a burst of neutrons. This device, which also used the
polonium-beryllium reaction, was given the name "urchin." It had
a radius of about one centimeter and fit into a space that had been
left in the otherwise solid pit.

In late February, the design of the Christy gadget was frozen
and a schedule set for actually building it in time for a summer
test. Oppenheimer created a high-level committee called the "Cowpuncher Committee" that was to ride herd on the entire enterprise.
But the previous spring he had created another committee, which
included Bainbridge, to find a site for a future test. They scouted several western states and finally came up with something in southern
New Mexico about two hundred miles from Los Alamos. It was a

part of the Alamogordo bombing range that was known as *Jornada del Muerto* – the Journey of Death – a name that had been given to it by the old Spanish wagon trains whose drivers often died in the desert here. It was a barren, isolated place far from any population center. Oppenheimer had to find a code name for the test and he chose "Trinity." When many years later he was asked by General Groves how he had come up with that name he said that, at the time, he had been reading the poetry of John Donne and had come across the line "Batter my heart, three person'd God." If anyone but Oppenheimer had given this explanation, one might be skeptical. But he had written poetry and had immersed himself in Donne. In September, the Alamogordo site was officially selected and, by November, work had been started on the construction of the necessary structures. By April, the detonators had been built and the production of the lenses was under way. By July the two hemispheres of the Christy gadget had been fabricated. On the 12th, the late Philip Morrison rode from Los Alamos to Alamogordo in an army sedan with the hemispheres in the back seat. When they were examined it turned out that there was a flaw that Smith was called on to repair. He wrote,

the hemispheres for the ... Trinity test were electro-plated and some aqueous electrolyte retained in a porous spot in one of them reacted and caused a tiny blister to form. This would have separated the mating surfaces enough to allow jetting during implosion and possible premature initiation. Postponement of the test was threatened, but I proposed the insertion of some rings of crinkled gold foil to prevent jetting and, late one night, I had the by-then-rare experience of working in the laboratory to make something with my own hands instead of watching someone else do it following instructions. This little blister made it necessary for me to become at the last moment a member of the team that was responsible for the final assembly of the first nuclear bomb. At approximately noon on 15 July 1945, at MacDonald's Ranch near Alamogordo in New Mexico, I put the proper amount of gold foil between the two hemispheres of plutonium. My fingers were the last to touch those portentous bits of warm metal. The feeling remains with me to this day, thirty-six years later.[3]

The next morning at 5:29:45 Mountain War Time the bomb exploded. It had a yield of approximately twenty kilotons. On August 9th a clone – "Fat Man" – was dropped on Nagasaki, ending the war. The people present at Trinity had different reactions. Rabi told me that he got goose flesh at the sight of the fire ball, realizing that something new and potentially terrible had just been born. Fermi tore a paper into strips to drop when the air blast reached him some forty seconds later. By seeing how they fell he could estimate the yield, which he found to be approximately ten kilotons. Bainbridge said to Oppenheimer shortly after the explosion, "Now we are all sons of bitches." And Oppenheimer? He died on February 18, 1967. A week later there was a memorial for him in Princeton. His brother Frank, whom I knew, was there. I knew that at the time of the explosion he had been standing next to Robert, so I took the opportunity of asking him what Oppenheimer had said. What he said was, "It worked."

[3] Some Recollections of Metallurgy at Los Alamos, 1943–1945, by Cyril Smith, J. Nuc. Matls., 100, 9 (1981).

9. Smoky and the Need to Know

Figure 22. Smoky – August 31, 1957, forty-four kilotons. Courtesy of the Department of Energy.

I RECEIVED MY PH.D. IN PHYSICS IN THE SPRING OF 1955. I HAD no job but, thanks largely to the help of the Harvard physicist and historian of science Gerald Holton, I found one. The Harvard cyclotron had a position that was known as the "house theorist." It had become vacant, and I applied for the job. Holton's recommendation sealed the deal. It was explained to me that one of my duties was to help the experimentalists with the theory of their experiments. But the experimentalists, people such as Norman Ramsey, knew vastly more about the theories of their experiments than I did. During the two years I had the job I believe that I answered one question from one experimentalist. The rest of the time I worked on my own problems and attended the seminars and colloquia at Harvard and MIT. I was asked to take over Ramsey's nuclear physics course for two weeks while he was in Washington testifying before Senator Joseph McCarthy, who was then on some anti-Harvard rampage. Ramsey did so well that McCarthy offered him a job. My job was due to end after two years so I applied to various places, including the Institute for Advanced Study. This was my first choice. Einstein and von Neumann had both died, but there was a glittering array of younger faculty members, including Freeman Dyson, who was a hero of mine because of his work on quantum electrodynamics, and T. D. Lee and C. N. Yang, whose sensational work on the "weak interactions" such as beta-decay was at the center of everyone's attention. And, of course, Oppenheimer was the director. I did not think that I had much chance to be accepted because I assumed that the competition would be fierce, but much to my amazement and delight, I learned in April that I had been accepted. The Institute term began in late September, which left open the summer in which to find an activity. I found one, which turned out to be somewhat more than I had bargained for.

This was the height of the Cold War and there were two weapons laboratories in full operation: Los Alamos and Livermore. Livermore, which had been founded in 1952 at the urging of Edward Teller, was located on a former naval base not far from Berkeley. Teller's complaint had been that Los Alamos was dragging its feet on the hydrogen bomb and that a competitive laboratory was needed. It turned out that all the essential work on the hydrogen bomb was done at Los Alamos after Teller left in a huff. As it happened, I had already met Teller. In the spring of 1954, someone from Harvard had recommended me as a possible recruit for Livermore. I did not yet have my degree and had no desire to work at a weapons laboratory, but I was curious to meet Teller. In late April there was a meeting of the American Physical Society in Washington, D.C., and I was given an appointment with Teller. We met in the lobby of the hotel where the Physical Society meeting was being held and where he had a suite. He suggested that we go to his suite, where we could talk more quietly. When we got there he explained that the next day he was giving a talk and the way he prepared was to find someone – me in this case – to rehearse with. I was instructed to interrupt him whenever I found anything wrong. He paced around the living room giving his talk. The little that I understood of it, I found very uninteresting. I had no comments. At one point he stopped pacing to say that he enjoyed talking physics rather than politics. I wish that I had asked him to elaborate. Only later did I learn that at that time he had given the devastating testimony at Oppenheimer's security hearing that caused him to lose his clearance. About *that* I would have had something to say. I never heard from Teller or Livermore again.

I have been searching my memory but I cannot remember how I was recruited to go to Los Alamos as an intern in the summer of 1957. Maybe Bainbridge, who was the department chairman and with whom I had an amicable relationship, recommended me. In any event, I got the offer subject to a security clearance. To work in the theory division at Los Alamos you needed what was known as a "Q-clearance" – one of the more rigorous FBI clearances. You had to write down all your dwelling places for years and years and God knows what else. I did not mention that I had a great aunt who subscribed to the *Daily Worker* and spoke frequently about the iniquities of the "bosses." I hope for their sake that she was not a member of the Party. Neighbors of mine told me that they had been interviewed. Years later, when the Freedom of Information Act was passed, I applied to get my FBI transcript. I got it after much delay due to the intervention of the late Senator Daniel Patrick Moynihan. By this time I was writing for the *New Yorker* and I thought I could write a lively piece that I was going to call "Friends and Neighbors" about all the dreadful things that had been said about me to the FBI. When the report came I was disappointed to see that it had been heavily redacted. Anything that looked at all interesting was covered up with black ink. No piece.

I also cannot remember how I got to Los Alamos. I did not have a car. Maybe I took the Topeka–Santa Fe train to Lamy, New Mexico. This is what people did during the war. They then checked into

109 East Palace Avenue in Santa Fe where, in an undistinguished-looking structure, they would encounter Dorothy McKibbin, who would make all the arrangements for them to get to Los Alamos. There was, and is, a small landing strip on the Los Alamos mesa. I doubt that I had enough rank to have gotten a flight in a small plane from the Albuquerque airport. Now there is shuttle bus service from the airport. Perhaps that is what I took. I wonder what I did when I got to the Los Alamos gate. At that time it was a closed city surrounded by guards and barbed wire. My guess is that it was even more closely guarded at that time than during the war. During the war no one outside Los Alamos knew what was going on there. Afterwards *everyone* knew. Did I just go up to the guard house and say, "I am Jeremy Bernstein and I am here for the summer?" I can't remember. I do remember that I was assigned to a room in a wooden dormitory that was left over from the war. I also acquired a bicycle, as the theoretical division was located quite far away from my dormitory. The head of the theoretical division at that time was a very droll Canadian named J. Carson Mark. He had come from the Canadian National Research Council to Los Alamos in 1945, and he never left. He retired in 1973 but continued to write very illuminating articles about things like nuclear proliferation until his death in 1997. I cannot believe that Carson would have given me any kind of orientation lecture. In fact, I had no assignment at all. But there was an equally young contemporary of mine named Ken Johnson who had also come to Los Alamos for the summer. We shared an office. I had thought of a nice problem in elementary particle physics, but I did not have the mathematical skills to solve it. Ken did, and we spent the summer working on it. We published a paper based on our work.[1] We gave Los Alamos as our return address. We must have cleared

---

[1] Decay of the Neutral π° Meson, *Phys. Rev.*, **109**, 189 (1958).

the paper with Carson. At the time we were both in transit, I to Princeton and Ken to the Bohr Institute in Copenhagen. The next fall Ken began a forty-year career at MIT. He died in 1999 at the age of 67.

As the summer proceeded, a number of more senior, distinguished, theoretical physicists appeared on the scene. They were mainly consulting on the project to make a controlled fusion reaction. In the next chapter, I discuss uncontrolled fusion reactions – the hydrogen bomb. But this was an attempt to design a machine that could use as its fuel light elements such as the isotopes of hydrogen to generate apparently limitless amounts of power. Heavy water, for example, could be extracted from the sea. Although much progress has been made in the last half century, it is still a work in progress. In 1957, it was a classified program. I recall trying to go to one of the seminars and being stopped at the door while my clearance was checked. A few years later it was declassified and it turned out that the Russians had a great deal to teach us. Among these consultants was a very brilliant theorist named Francis Low. He had done some work that I much admired and we became friends. One of the things we shared was a passion for tennis. We played innumerable sets of singles and doubles and even represented Los Alamos in a match against Santa Fe. The Santa Fe players must have entered the site under escort. Toward the middle of August, Francis announced that he would be away for a few days. When I asked where he was going he said to Mercury, Nevada, to watch some atomic bomb tests. This caught me by surprise.

I, of course, knew that the primary mission of Los Alamos at that time was the design and testing of nuclear weapons. But this was being done by people in other divisions. There were no colloquia – as there had been during the war – in which all the cleared people in the laboratory heard about nuclear weapons. I had no idea that there was anything like Serber's primer. In 1957, there were still people there who had been there during the war. Carson was one, along with the aforementioned James Tuck. I have an ineluctable memory of the Falstaffian Tuck at one of the afternoon division teas – which I imagine must have come to Los Alamos with the British delegation – looking out at us from his chair and announcing in his redolent English accent, "The days of the great Los Alamos teas are passed." He must have had in mind the teas at which the likes of Fermi, Bohr, von Neumann, and Oppenheimer were present. I am sure that Tuck and Carson were working on weapons that summer. But one rule with the Q-clearance, which had been in effect since the war, was "the need to know." If you needed to know you could be told, otherwise it was just as well not to ask. Nothing that I was doing with Ken gave me the need to know. I do recall on one of my exploratory bicycle rides going down a road that led to a cave that was being guarded by people with machine guns. I beat a hasty retreat after explaining that I had gotten lost. I had no idea that Francis had been working on weapons. He told me that he hadn't but that Carson had invited him to watch the tests and that his curiosity had gotten the better of him. I felt decidedly jealous. Francis said that if I asked Carson maybe he would let me come as well. I asked and Carson said it would be fine with him provided that I paid my way – airfares, lodging, and the rest. I readily agreed. Thus it was that on the afternoon of August 30th Carson, Francis, and I departed Los Alamos in a small plane, first to Albuquerque, and then by commercial airline to Las Vegas. Mercury is about sixty-five miles northwest of the city – pretty close to an atomic bomb aboveground test range.

To explain what happened next I have to back up some. What I am going to tell you may sound frivolous in light of the context, but I think it was part of the mind-set. Sometime prior to my visit to Los Alamos a mathematician had published a paper about casino blackjack as it was then played. It created something of a sensation because he showed that if you followed an optimum strategy you could beat the house. As I recall, you had to do a little card counting, which was possible because the dealer used only one or two decks. The last time I played casino blackjack was in an Indian casino off the road from Santa Fe to Los Alamos. I noticed that now they are using many decks so that card counting is essentially impossible. In any event, so many people from the laboratory were going to Las Vegas that the strategy was tested at Los Alamos prior to the trip. They ran millions of hands on the big Los Alamos computer using the mathematician's strategy. The result was confirmed, and those of us going to Mercury were given a little paper folio with the strategy on it. Francis, who was more quantitative than I am, calculated how much you could actually earn if you played according to the strategy for a year. I think he concluded that it would be less than the minimum wage in a state like Alabama. Thus, after we landed in Las Vegas in early evening, we went directly to one of the casinos. This one must have had quite a Mercury clientele because they had a signal system – a blue light – that told people that the test the following morning was on. The tests were frequently postponed because of the weather. Most of the tests were ground bursts in which the fireball hits the ground – as opposed to air bursts where it doesn't – so local radioactive fallout was a real concern. After an hour or two of blackjack the blue light went on and we all departed for Mercury by automobile.

The test that was scheduled had the code name "Smoky." All the Livermore tests in 1957 had names associated with North American mountains. There was Rainier, Hood, Shasta, Owens, and the like.

So this test was a Livermore test. The Los Alamos tests were named after scientists like Kepler, Galileo, and Boltzmann. Smoky was scheduled for 5:00 the following morning. The tests at Mercury were all at sunrise so that it would be light enough for aerial photography but dark enough so that the visual effects would be the clearest. Carson got us up a little after 4:00. He took us to a room where the site meteorologist was plotting things like air currents. He did not like what he saw, so the test was postponed for half an hour. Then we went outside. I am not sure exactly how far we were from the seven-hundred-foot tower on the top of which Smoky was placed. It turned out that this was the tallest tower ever built for an American test, about two-thirds the height of the Empire State Building (see Figure 23). I am sure that we were at least five miles away. I also did not realize that this test location had been specially selected because the terrain was rolling and they wanted to see how this affected the explosion. Behind the tower was a high hill covered with yucca – that ubiquitous flowering plant of the desert southwest. Some varieties are called Joshua trees. They look like penitents. The Joshua tree–covered hill made a kind of natural amphitheater. Carson gave us a very short briefing. About all he said was that at the time of the explosion we were to look away and count to ten – otherwise the light could blind us. Then we could turn around.

There was a voice from a loudspeaker that counted down from ten. I think it had been preceded by an alarm that warned everyone that there was about to be a test. As instructed, I turned my back to the tower. At "zero" there was an almost blinding reflected light. I counted to ten and turned around. No words adequately convey the vision of that infernal fireball. No picture that I have ever seen of Smoky shows what for me was the most vivid visual impression. The entire mountain of Joshua trees was on fire. The plants were blazing like some sort of monstrous auto-da-fé. There was also the size of the fireball. It dominated the horizon. There was no noise. The explosion had been totally silent. Then some seconds later there came the shock wave. It made a slightly painful cracking sound in my ears. This was followed by a rush of wind. At Trinity, Kistiakowski was knocked over by this wind rush, although he was five miles away from the explosion. Smoky was more than twice as powerful as the "gadget," so we must have been farther away, since I have no recollection of being pushed around by the wind. Then came the noise of the explosion – rolling thunder. At Trinity the sole representative of the press was the doyen of science journalists, William L. Laurence of the *New York Times*. Feynman had been put in charge of looking after him. When, after the shock wave, the noise came, Laurence asked, "What was that?" This is often cited to show what a scientific illiterate Laurence was. But actually it is not a bad question as I will later explain. The fireball dimmed and was replaced by a rising funnel cloud that looked purple and black. There was something dreadfully menacing about that cloud. It was then that I remembered what Oppenheimer had thought, but apparently had not said, at the time of the Trinity explosion. In the 1930s he had studied Sanskrit in Berkeley and had read the Bhagavad Gita. This verse had stayed with him:

> If the radiance of a thousand suns
> Were to burst at once into the sky,
> That would be like the splendor of the Mighty One…
> I am become Death,
> The shatterer of Worlds.

That menacing, dreadful, hovering radioactive cloud had become for me a symbol of death.

After the explosion we did not have much to say to one another. We went back to our bunkhouse for a little more sleep. A little later I was wakened by the sound of helicopters. I had no idea what that meant. "They're flying and flying," Carson said laconically. What it meant I will explain, along with the full significance of this test, after I describe the rest of our experience. After lunch Carson took us on an automobile tour of Mercury. Where previous tests had been, the ground was green and vitrified. The desert soil had been turned into glass. There were signs here and there warning of high levels of radioactivity. We drove quickly through these areas with the windows closed. Carson stopped at a five-hundred-foot-high tower that looked much like the Smoky tower in Figure 23. There was an open lift on the outside like the one you see in the picture. One could make out a platform on top. Carson explained that this was the tower for a Los Alamos test that was called "Galileo." It was scheduled for two days later and the bomb was being prepared on the platform. We got into the lift. Where it stopped there was a rather flimsy metal ladder that led to the platform. When I got on the ladder and looked down, almost five hundred feet to the desert, I had a momentary feeling of acrophobia that was more frightening to me than the thought of the bomb on the platform. On the platform was the bomb, being worked over by a half-dozen men. It was much larger than I had expected. The actual pit was buried somewhere in the middle of the device. There was a curious clicking sound from a pump. From my cyclotron days I thought that it sounded cryogenic, although I could not imagine what it was doing here. Since I did not need to know, I did not ask. As it turned out, and as I explain more carefully in the next chapter, neither Smoky nor Galileo were conventional fission bombs. They were what might better be described as two-thirds of a hydrogen bomb. The cryogenics was connected to that.

Figure 23. Smoky's tower. The bomb was on the platform at the top. The tower was seven hundred feet tall. Photo courtesy of National Nuclear Security Administration/Nevada Site Office.

I guess we stayed on the platform for something like a half hour. Carson had some matters to discuss. Then we descended the ladder and the lift and took off again in the car. Our last stop was a nondescript-looking stone blockhouse located at a far end of the site. Carson offered no explanation. He opened the door and when I went inside I had almost the instinct to run away. Its walls were lined with shelves. On the shelves, neatly arranged, were pits of unexploded nuclear weapons. I think there were several dozen – enough to blow up a continent. Carson must have seen my reaction because without saying anything he gave me one to hold. I think he said, "Don't drop it." It was about the size and weight of a bowling ball. Although it did not occur to me then, the sphere must have been hollow because a solid sphere of plutonium of that size would have required a derrick to lift. It was warm to the touch. The alpha-particle decay of the plutonium had heated the stainless steel cladding encapsulating the plutonium. There were no explosives on the sphere; they would have fitted above the cladding. But there was a man at a bench in the process of gluing – I think it was gluing – explosive lenses to one of the pits. He seemed completely nonchalant as he went about this. What I have never forgotten was that next to him was a woman seated on a chair, knitting. I suppose that she was his wife – knitting, simply knitting silently while her husband was gluing high explosives to the pit of an atomic bomb. It is odd how memory works. For many years Francis insisted that this woman was a figment of my imagination. Carson did not remember her either. But a few years ago Francis unexpectedly called me to say that he now remembered her just as distinctly as I did. A day later we watched Galileo explode, evaporating the tower.

I wish I could tell you that this experience made me wise. It actually made me foolish. At the end of this chapter I give an example that, after almost a half century, still makes me cringe. Although after this time in Mercury I still knew next to nothing about nuclear weapons, I somehow felt that I had crossed the divide into the secret world. I had actually held in my hands the motive center of a nuclear weapon and I thought that this, in itself, had given me some kind of power. I almost felt sorry for people who had not had this experience. They were not part of the secret. It took some time before I realized how totally absurd this was. But it gave me an understanding of how corrupting this kind of power can be. It took me an even longer time to finally understand what I had seen in Mercury in those few days. I really did not put it all together until I began writing this book. That is what I will now explain.

Smoky and Galileo were part of the longest and most extensive test series ever done at Mercury. It was called "Operation Plumbbob." It consisted of twenty-nine detonations, beginning with Boltzmann on the 28th of May and ending with Morgan on the 7th of October. Some of the tests were on towers, others from balloons, one in a shaft, and another in a tunnel. In a "boys will be boys" gesture the one in the shaft had an odd twist that resembled what we used to do with firecrackers when we were kids. We'd put them in a tin can to see what happened when they went off. Here the grown-up boys put a lid on the shaft to see what would happen. They figured that so much energy would be imparted to the lid that it would reach a velocity greater than that needed to escape the gravitation of the Earth. It would become, before *Sputnik*, the first space vehicle. It is not likely that this happened, but the lid was never found after the explosion. The Plumbbob explosions ranged in yield from half a ton – Lassen – to seventy-four kilotons – Hood. They released 58,300 kilocuries of radio iodine into the atmosphere. A "curie" is a unit of radioactive disintegration that is equivalent to thirty-seven billion decays of anything per second. To get some idea of scale, one thousandth of a curie is about what would be used in a liver scan. The amount of radio iodine that Plumbbob put into the atmosphere was estimated to have caused an additional 38,000 cases of thyroid cancer, leading to some 2,000 deaths. Most of this radiation was deposited in the northeastern, far western, and midwestern states. There was a cluster in Maine. But this is not the only thing that made the Plumbbob tests notorious. The Department of Defense decided that this would be a splendid opportunity to study the reactions of soldiers and marines to atomic warfare. During the series some 18,000 servicemen participated. Because the tests were frequently postponed they were able to make substantial contributions to the Las Vegas economy at all levels. At Smoky, some three thousand soldiers were brought close to ground zero not long after the explosion. They had watched the explosion from trenches about a mile away. This explained the sound of helicopters that I had heard that morning. It is difficult to imagine now that our defense establishment would have done something so absurd. But there it was. The health of the Smoky soldiers was followed for several decades. In 1980, a survey showed that their leukemia rates were elevated. Four would have been a baseline number of cases expected in the general population, whereas there were ten.

I now want to turn to the science of what I saw, or didn't see. Much of it happened too fast to see. In this respect, tribute should be paid to the electrical engineer Harold Eugene Edgerton. Edgerton was born in 1903. He grew up in Nebraska and went to MIT for his doctorate. His thesis was on the use of stroboscopes to study motors. He then began using stroboscopes to photograph things like speeding bullets and hummingbirds. For many years he collaborated with the *Life* magazine photographer Gjon Mili. In the summer of 1945, to his everlasting regret, Edgerton was persuaded to use his techniques to film the Trinity test. He had a camera with about 3,500 feet of 35-millimeter film that shot the sequence in less than millisecond intervals. The camera was going so fast that after shooting about two thirds of a mile of film it exploded. Figure 24 shows a sequence that begins at a tenth of a millisecond after the explosion. Our job is to understand what it means. In this I am going to restrict myself to a pure fission weapon. As I mentioned, both Smoky and Galileo involved nuclear fusion as well. This is the source of energy that I discuss in the next chapter.

On the time scales of these explosions, Professor Edgerton's photographs start somewhat late. The nuclear fission is over in microseconds. It is interesting to note that most of the fission energy is produced in the last few generations of the chain reaction, when the largest number of neutrons have gotten into the act. The fission of the plutonium proceeds until the sphere has expanded from the gadget's compressed pit radius to second criticality, when more neutrons are escaping than causing fission. As I have discussed, the time this takes is one of the important things that determines the efficiency of the weapon. In the case of the plutonium Nagasaki bomb, Fat Man, of the approximately twelve pounds of plutonium they began with, about two pounds were fissioned. The rest drifted off into the atmosphere. The fission energy is provided by the mass difference between the initial and final nuclei using Einstein's $E = mc^2$. Most of this energy is transferred to the energy of motion – the "kinetic energy" – of the fission fragments. The extreme temperatures vaporize all the solid material. This produces a gas of highly agitated fission fragments, as well as the debris from the cladding, the tamper, and the rest. The macroscopic manifestation of this agitation is heat. Depending on the type of bomb, this temperature is initially

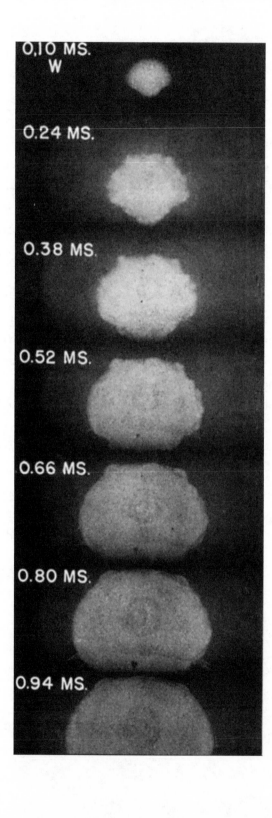

at least sixty million to one hundred million degrees centigrade. For comparison, the surface temperature of the Sun is six thousand degrees centigrade. The central temperature is about sixteen million degrees. These heated bodies radiate electromagnetic energy. A spectrum of radiation is emitted at different wavelengths. For this kind of radiation the spectrum is independent of the nature of the body and depends fundamentally only on the temperature. The hotter the body, the greater the intensity of the radiation at the shorter wavelengths. One can ask at what wavelength, at a given temperature, the intensity peaks. At six thousand degrees it peaks at about 5×10^{-7} meter. This is somewhere in the visible range for light. If you are a creationist you can say that this is because the eye was created along with the Sun. If you are an evolutionist you can say that the eye evolved to process the light from the Sun in the most useful way. At sixty million degrees – a typical bomb temperature – the radiation peaks at about ten thousand times smaller wavelengths, which puts you in the X-ray regime. The conclusion is that most of the initial energy from the bomb is converted into X-rays. There is also some energy in the electrons that, at these temperatures, have been torn off the atoms, and some in gamma radiation – high-energy electromagnetic radiation – that is produced in the decay of the radioactive fission isotopes.

The mention of electrons in connection with the bomb has tempted me into a diversion. Do not worry, I will not lose my train of thought. This is another drama of nuclear folly with several actors. I played, as you will see, a very minor role. The first actor on the scene was the late James Van Allen, who died in August 2006. He was born in Iowa in 1914 and attended school there, finally taking his

Figure 24. Edgerton's photographs of the Alamogordo test. "MS" stands for thousandths of a second. Courtesy of Los Alamos National Laboratory.

Ph.D. from the University of Iowa in 1939. After being discharged from the navy, in which he served during the Second World War, he became interested in high-altitude experiments using captured German V-2 rockets. He soon joined a collaboration to develop a kind of rocket that they called a "Rockoon." The rockets were launched from high-altitude balloons that reached the fringe of the atmosphere. Van Allen was certainly not the first person to suggest that electrically charged particles might get trapped above the atmosphere in the magnetic lines of force that emanate from the Earth's magnetic field, but he was the first person to design and carry out experiments to find them. Indeed, the Rockoons already gave some preliminary indication that these particles were there. This brings us to 1957, which was part of the International Geophysical Year – actually a period of eighteen months. That year, *Sputnik* was launched into orbit by the Russians, and the following year, using a Jupiter missile, our satellite, *Explorer I*, was launched into orbit. Using this satellite, the so-called inner Van Allen belt was discovered. It lies about 9,400 kilometers above the equator. Its main content is protons that are thought to have been produced in the beta decays of the neutrons originating with cosmic rays. In 1958, *Pioneer* 3, which was supposed to be a lunar probe, was launched by a Juno rocket. It only reached an altitude of 63,000 miles, but it discovered a second belt composed largely of electrons. As a result, Van Allen was on the cover of *Time* magazine in 1959. However, he now moves off our stage and is replaced by an even more colorful figure, Nicholas Christofilos.

As the name suggests, he was Greek – he died in 1972 – but he was born in Boston in 1916. The family moved back to Greece and Christofilos was educated in Athens, graduating in 1938 from the National Polytechnic with an electrical engineering degree. He then went to work for an elevator maintenance firm and later established his own. During his spare time, he read physics books and, after the war, he used the United States Information Services Library in Athens to read journals like the *Physical Review*. In 1946, he invented a scheme for accelerating particles, which he patented in the United States and Greece, sending copies to Berkeley. This scheme had shortcomings, but in 1949 he invented another, also patented and sent to Berkeley. The patent was granted in 1956. It was independently rediscovered in the United States by a team at the Brookhaven National Laboratory on Long Island, New York. It became the basis of the next generation of particle accelerators, and Christofilos, who had shown up at Brookhaven to insist on his priority, was offered, and accepted, a job there. He was, it appears, something of a handful. In 1956, he changed jobs and went to Livermore, where he enters our story.

By all accounts, Christofilos embraced the national defense side of the laboratory with characteristic gusto. In 1957, prior to Van Allen's discovery, he had an epiphany. He realized that if electrons got caught in these high-altitude magnetic field lines then, because of the shape of the field lines, the electrons would be reflected back and forth – something that was called the "magnetic mirror" effect. To Christofilos the implication was clear. There might be bands of electrons in space and, even better, if there weren't, he knew how to make them. What you would do is to put an atomic bomb on a rocket, send it above the atmosphere, and explode it, thus producing electrons that would be caught in an artificial belt. Under most circumstances this idea was one that you might find on the back of a cereal box, with other "magic" devices. But Christofilos also saw national defense implications. Such a band, he decided, might intercept ballistic missiles. He apparently conveyed this to Teller, who shared his enthusiasm. The idea was a primal form of Ronald Reagan's Star Wars. But Teller was in a

position to bring this chimera about. The project Argus was created. Argus, you may recall, was the giant of Greek mythology
with innumerable eyes that never closed; the perfect guardian. During the period from August 27 to September 6, 1958, three small
nuclear bombs – in the one- to two-kiloton range – were launched.
They were exploded over the South Atlantic at altitudes of about
one hundred eighty kilometers. The effects were measured by an
Explorer satellite and, indeed, artificial Van Allen belts were produced. An aurora appeared over Gibraltar – a phenomenon never
seen there before or since. The belts proved too evanescent to play
any role in national defense, however.

My role in all of this was peripheral but, if I may say so, enter-
taining. In the summer of 1958, I got a consulting job at the RAND
Corporation in Santa Monica. I soon discovered that they had as
little idea of what I should do as Los Alamos had had the previous
summer. I spent the time doing my own work, playing tennis, and
going to the beach. At one point, Herman Kahn, who was in the
physics division, asked me to read and comment on a draft of what
became his book, *On Thermonuclear War*. This is the book in which
Kahn explains that, although we would suffer a few mega-deaths in
a nuclear exchange, after it was over, if we played our cards right, we
could carry on pretty much as before. The manuscript appalled me,
but I thought that getting into an argument with Kahn would serve
no purpose. In the period when Kubrick was gestating the material that became *Dr. Strangelove*, he immersed himself in RAND
Corporation reports, including those of Kahn. In the picture, the
RAND Corporation became the BLAND Corporation, and some of
the dialogue comes directly from Kahn, who was certainly a model
for Strangelove. Peter Sellers' accent was taken from the photographer Weegee – née Arthur Fellig – who was on the set. In any event,

my tranquil existence at RAND was interrupted by what appeared to be an explosion of activity in the physics division. I think there had been a visit by Teller or Christofilos, or both. It appeared as if the British radio astronomer Sir Bernard Lovell, who had somehow gotten wind of the impending Argus explosions, was making angry noises. Lovell was concerned that they would produce these artificial belts before they had a decent chance to study the real ones and that the electrons running around as a result might interfere with radio astronomy. Our division at RAND had been mobilized to demonstrate theoretically that none of these concerns were serious. At least I think that is what we had been mobilized for. Nothing was explained to me. However, one of the Latter brothers – Richard and Albert, physicists who ran the division – appeared in my office with a very long list of numbers that I was instructed to add up. That was that.

Now, in the fall of 1960, I had a job at Brookhaven. To explain what happened I need to explain something about Brookhaven. Whereas, like Los Alamos, it ran on government funds, it was managed by a consortium of universities. Los Alamos was, and is, managed by the University of California. As far as I know there was almost no classified work being done at Brookhaven; nonetheless, the facility was guarded. You needed some sort of minor clearance to work there. My Q-clearance had lapsed. One afternoon the phone rang in my office. It was some security officer. They had just received a classified package from RAND. They wanted my permission to open it. I said that I would not give them permission without knowing the contents of the package. They said they could not tell me because it was classified. We were at a stalemate. To resolve matters they called RAND. They were informed that it was a report on upper atmospheric nuclear tests. That is all they were willing to say except that I was one of the authors. This came as a surprise because all I had done was to add up a column of numbers. Be that as it may, because my Q-clearance had lapsed, the package was sent back to RAND unopened. I never found out what was in the report.

Not to be outdone, in 1962, Los Alamos supplied a hydrogen bomb to be detonated in outer space. As we will see in the next chapter, such weapons have yields equivalent to millions of tons – megatons – of TNT. The Los Alamos bomb was put on a Thor rocket and launched on July 9th above Johnson Island in the Pacific – project Starfish. It exploded at an altitude of 250 miles with a yield of 1.4 megatons. The result was more than was bargained for. An electron belt was temporarily created that managed to destroy seven satellites, including the first commercial communications satellite.[2] Now we can return to our discussion of nuclear explosion.

Up to the production of X-rays, all the steps have been independent of the environment in which the bomb explodes. The same steps would occur whether the bomb was exploded in the vacuum of outer space or at the Earth's surface. In the next steps the environment plays a crucial role. I will focus on the sort of ground-level bursts that I witnessed. The essential point is that these take place in the Earth's atmosphere – the "troposphere" – which is made up mainly of nitrogen and oxygen. The primal X-rays are absorbed within a few feet and heat up the surrounding air. This is the origin of the "fireball"; Edgerton's photographs show how it develops, millisecond by millisecond. The initial temperatures are tens of millions of degrees, with a surface brightness trillions of times greater than

---

[2] I am grateful to Richard Garwin for a communication in which he informed me that he was in charge of a government commission to attempt to determine why this low-altitude blast injected electrons into the belt when it was not supposed to.

the Sun.[3] There is a famous story about the Trinity test. A family was taking their blind daughter back to school when the explosion occurred; she saw the flash of light. The next step in the process is rather complicated. It occurs when the fireball is a few meters across, at something like a tenth of a millisecond for a twenty-kiloton bomb. Because of the expansion, the temperature drops to something like 300,000 degrees centigrade. Because of the immense heat, the pressure inside the fireball is a few thousand times greater than the ambient air pressure. The air near the surface of the fireball is very sharply compressed by this pressure difference, which produces a shock wave – or "blast wave." This shock wave moves much faster than the speed of sound in air and, indeed, begins to heat up the air it comes in contact with, to temperatures approaching 30,000 degrees centigrade. This is hot enough so that some of the electrons are stripped from the air molecules and the air itself becomes incandescent. This ionized air absorbs the much hotter radiation from the interior of the fireball. The interior becomes invisible to an outside observer. Edgerton's photographs, at this stage, show the incandescent air external to the fireball. At about fifteen milliseconds for a bomb like the gadget – twenty kilotons – the temperature of the incandescent air drops to the point at which it is no longer opaque to the radiation from the fireball. The interior of the fireball is still at a temperature greater than the surface temperature of the Sun. Edgerton's photographs after fifteen milliseconds show this interior. For example, the photograph in Figure 25 shows the interior of the fireball at sixteen hundredths of a second. It takes tens of milliseconds for the eye to transmit the visual signal to the brain. The first thing

[3] A scientifically oriented reader may wonder why it is only trillions. There are other effects absorbing the radiation. If this was not the case, the intensity would be proportional to the fourth power of the temperature.

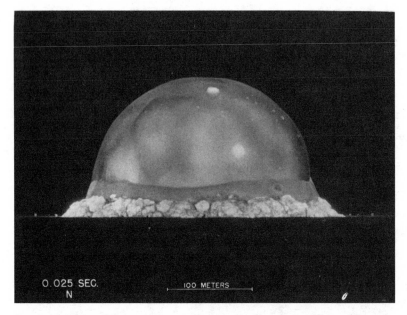

Figure 25. The gadget at sixteen milliseconds after the explosion. Note the width scale. One hundred meters is about the length of a football field. Courtesy of Los Alamos National Laboratory.

one would see after the explosion, if one had been foolish enough to look, would be a very intense bluish white light. The photograph of Smoky in Figure 22 at the beginning of this chapter is also of the interior of the fireball, but the timing is somewhat different because Smoky was a more powerful explosion. At this stage of the explosion, the shock front is still supersonic. By the way, this pair of luminous flashes – the original incandescence of the air due to the X-rays and the release of the radiation from the interior of the fireball – separated by milliseconds, is characteristic of nuclear explosions. It is a signature that is used by spy satellites to detect them.

I would like to mention an additional thing about the shock wave. By definition a "surface blast" is one in which the fireball hits the ground. In "air blasts," like those of Nagasaki and Hiroshima, the fireball does not hit the ground. If the fireball hits the ground there are two significant effects. Ground material is scooped up and made radioactive. I will come back to this in a moment. But also the shock wave at the surface of the fireball can be reflected back into the interior. Because the interior is so much hotter, this reflected shock wave moves faster than the original one. It catches up to it and the two merge.

So now let us review the sequence of events visible and audible to an observer. First there is the initial flash of light. If one has any choice in the matter, one should not look at this, nor at the subsequent incandescent air. At Smoky, when I turned around at the count of ten, I saw the interior light from the fireball as depicted in the picture at the beginning of this chapter. The differences in brightness across the fireball presumably reflect differences in density of the internal material. The next thing I experienced was the somewhat unpleasant click of the shock wave in my ears, followed by the wind. And only then did I hear the noise from the explosion. So when William Laurence, witnessing Trinity next to Feynman, upon hearing the noise, asked "What's that?" he was actually asking a rather good question. I wonder what Feynman told him.

The reader may be wondering where in all of this is the mushroom-shaped cloud that is so emblematic of the nuclear age. First, let me point out that it is not unique to nuclear explosions, but it is dramatic in these explosions because of the magnitudes involved. In the nuclear explosion the fireball rises like a hot-air balloon. The air below it is much cooler and is sucked into the fireball. This creates a kind of shaft beneath the fireball that contains, in addition to air, debris, which can be very radioactive, that is sucked up from the ground. This shaft can be very dark. Once the cool air

Figure 26. The formation of a mushroom cloud. Note the flow of gases.

gets inside the fireball it begins to circulate as shown in the diagram in Figure 26. This can then distort the fireball into elongated shapes. It is the elongated fireball on top of the "stem" of cool air that resembles a mushroom. Water drops are extracted from the cooler air that has been drawn up into the fireball. These droplets form clouds. The Nagasaki photograph in Figure 27 shows these various effects. I do not have any recollection of seeing a white mushroom cloud when I witnessed Smoky. I think it was above the stem that I did see that dark and deadly radioactive cloud.

The subject of the effects of these weapons is complex and important. I will just do a very brief overview here and then come back to it in the next chapter. Here I will discuss the Hiroshima and Nagasaki types of fission bombs, which were in the twenty-kiloton regime. The first form of damage that I will mention is from thermal radiation – heat. With the enormous temperatures involved one would expect this to be significant. In fact, for a twenty-kiloton

Figure 27. The mushroom cloud formed by the Nagasaki bomb. Note the dark column below the cloud. Courtesy of the National Archives.

bomb, anyone who is not protected and is within about a mile and a half of ground zero will receive third-degree burns, which are almost certainly fatal. With a one-megaton hydrogen bomb this region is extended to some seven miles. With a twenty-kiloton explosion anyone up to a distance of about two and a half miles will receive first-degree burns. With a megaton bomb this region extends to more than ten miles. Smoky had a yield of more than twice that of the gadget but I do not recall any thermal effects. I think we were – deliberately – too far away. The next source of damage I want to mention is the blast wave. This damage is hard to quantify because much depends on the kinds of structures involved. Recall that at five miles away, Kistiakowski was knocked down by the blast wave from the gadget. At one thousand meters from the blast center, the winds were as strong as the strongest measured typhoon. At Hiroshima, the collapsed buildings became tinder for the fires started largely by overturned cooking stoves and a firestorm was created. At Nagasaki there was no true firestorm, but there were intense fires at many locations, which lasted for several hours. This sort of damage is common to any high explosive but what is unique to atomic weapons is the radiation. This radiation occurs at different times. First there is the "prompt" radiation, which arrives soon after the explosion and consists of neutrons, gamma rays, and electrons. Of these the most damaging are the neutrons. After the prompt radiation there is the fallout. This may begin an hour or two after the explosion and last for a day or so. Then there can be accumulated low-level radiation to which someone remaining in a contaminated area can be exposed for years. All of this radiation can have devastating effects ranging from immediate manifestations of radiation sickness to the latent development of cancers. As I mentioned at the beginning of the book, by December 15, 1945, something like ninety thousand people had died as a result of the Hiroshima bomb.

I want very briefly to return to the comparison with Timothy McVeigh's Ryder truck. It held 2.5 tons of high explosives. Thus one kiloton is equivalent to 400 Ryder trucks. The Nagasaki bomb was equivalent to 8,000 Ryder trucks, and a hydrogen bomb of twenty megatons would be equivalent to 8,000,000 Ryder trucks. These figures speak for themselves.

I left Los Alamos not long after Francis and I got back from Mercury. He was driving east to take up his new job at MIT. I was going to Princeton but I wanted to stop off in Lake Forest, near Chicago, to see a girlfriend, so Francis and I drove, in his car, as far as Illinois. I don't recall that we ever discussed the bomb or the tests we had seen. Not having a need to know, I did not ask any questions at Los Alamos either. I was extremely ignorant about what I had seen. But, as I said earlier, it had given me an absurd sense of superiority. It was as if I had crossed over into the secret world. As part of this I had gotten the idea that these aboveground nuclear tests were both necessary and important. If someone had asked me why, I would have had no plausible explanation. I mention this because of something that happened on that visit to Lake Forest. To understand it you need to recall that in 1956 Adlai Stevenson lost the presidential election to Dwight Eisenhower. Stevenson had made a nuclear test ban treaty one of the elements of his campaign. At one point, for example, he said, "there is not peace – real peace – while more than half of our federal budget goes in an armament race ... and the earth's atmosphere is contaminated from week to week by exploding hydrogen bombs."[4] He was attacked by Eisenhower, Nixon, and all their minions, and one of the issues that Eisenhower won the election on was maintaining the nuclear tests. It was a bitter pill

---

[4] See, for example, lewrockwell.com/wittner/wittner21.html, where this quote is given along with a discussion of the campaign.

for Stevenson. Now you will be able to understand my experience. My girlfriend's family were neighbors and close friends of Governor Stevenson. I told her mother a little about watching the tests and how I was persuaded of their importance. She said that that very afternoon we were going to a cocktail party and that I would have the chance to tell Governor Stevenson myself. Stevenson was one of my heroes, and I was eager to meet him. Left to my own devices I probably would not have said anything about the tests. But the first thing that my girlfriend's mother said after introducing me to Stevenson was that I had just come from Los Alamos and had seen some tests and had concluded that they were important. Stevenson gave me a look of contempt, turned his back, and walked away without saying another word. I have never forgotten that moment.

The great irony is that, faced with growing world pressure, the following year Eisenhower began negotiating with the Russians for a test ban treaty. By 1960, every major presidential candidate had endorsed such a treaty. After Kennedy won the election, he appointed Stevenson ambassador to the United Nations. He was in Moscow on August 5, 1963, when Khrushchev signed the partial test ban treaty in which we and the Russians agreed to stop nuclear testing in the atmosphere (see Figure 28). By 1962, our total tests had produced a yield of 153.8 megatons, while the Russians had produced 281.6 megatons. One can only imagine what Stevenson must have been thinking when he was watching Khrushchev. He died two years later in London.

Since 1963, a Comprehensive Test Ban Treaty with strong United Nations support was drafted; it became available for signing in 1996. If it is ever ratified by the signatories it would ban all nuclear tests of any kind. India and Pakistan did not sign and neither did North Korea. Neither Iran nor the United States have ratified it. Neither

Figure 28. At the signing of the partial test ban treaty. Stevenson is just to Khrushchev's right. Lawrence Berkeley National Lab – Glen Seaborg.

has Israel, China, Egypt, or the Marshall Islands. The Russians have, along with 137 other countries. Until several of the countries who have not ratified it do so, the treaty will not go into effect. There seems to be no sense of urgency for these countries to ratify, so the nuclear arms race goes on.

10. Fusion

Figure 29. Edward Teller (left) and Andrei Sakharov (right) in 1988. ©
Bettmann/CORBIS.

O PPENHEIMER ONCE REMARKED THAT THE ONLY PERSON HE knew who was identical to his own caricature was the physicist Wolfgang Pauli. Whereas I agree that Pauli was identical to his caricature, I think that Oppenheimer overlooked George Gamow. Gamow, who died in 1968, was an outsized (he was 6'3"), exuberant Russian eccentric. He was also one of the most gifted and imaginative physicists of the twentieth century. He was born in Odessa in 1904 and attended Odessa's Novrossia University and then, from 1922 to 1928, the University of Leningrad. In 1928 he enrolled in a summer school in Göttingen. This was just at the beginning of the quantum theory revolution and Göttingen was one of the nodes. Max Born was a professor there, and Heisenberg had been his assistant when he made the initial breakthrough. Born gave the first probabilistic interpretation of the theory. He had several very outstanding students, including Oppenheimer. That summer, Gamow made his first important contribution to applications of the quantum theory. I will return to it shortly, as it plays a role in our story. Bohr was sufficiently impressed to invite Gamow to his institute in Copenhagen, where he spent a year. During this time he invented the liquid-drop model of the nucleus. Not much attention was paid to it until the 1930s, when Bohr embraced it as a model for nuclear reactions, including fission. Gamow then spent a year in Cambridge under Rutherford's wing, after which he returned to the Soviet Union to become a professor at the University of Leningrad. Starting in 1932, he attempted to escape several times. His first attempt, which he made with his new wife, was by kayak. They took with them paddles, eggs, chocolate, strawberries, and two bottles of brandy. Their idea was to make the 170-mile crossing of the Black Sea to Turkey. After a day, bad weather forced them back to Russia, where Gamow explained that he had only been giving the kayak a sea trial. There was a conference in Brussels in 1933 to which Gamow was invited. He took his wife and never returned to Russia. He got a professorship at George Washington University in Washington, D.C. One of his conditions was that they allow him to bring Edward Teller, who was then in England. Until the war, he and Teller had a very fruitful collaboration. It is sometimes said that Gamow was not asked to come to Los Alamos because his Russian background would have made a security clearance difficult. I think that General Groves would have cleared Attila the Hun if he had had good ideas about implosion. My guess is that Oppenheimer decided that he did not need another prankster. Feynman was enough.

During Gamow's collaborative period with Teller they worked on applications of nuclear physics to stellar energy. This gave Gamow the idea of studying the physics of the universe under the assumption that it began with a cosmic explosion. In various collaborations he published papers, largely ignored, in which he spelled out how some of the light elements would have been produced. Part of the reason, I think, that Gamow's papers did not receive the attention they deserved was because, at the time he wrote them, there was no experimental basis for testing their assertions. There was a rival theory championed by the British astronomer Fred Hoyle that claimed that there was no origin, but that matter in small amounts was being continuously created from the vacuum. Since neither Gamow nor Hoyle had any data to support their ideas, they defended them by insulting each other. Hoyle's big insult was to refer to Gamow's model derisively as the "Big Bang." Hoyle lived long enough to see this become the term of art. There are many stories about Gamow but one that I think is true was told to me by someone who was involved. There was a meeting of astronomers to which my colleague and Gamow were both invited. My colleague noticed Gamow in the hotel bar taking his daily quaff. He bribed a waitress to go up to Gamow and say, "There is a telephone call for you Professor Hoyle." Without missing a beat, Gamow said, "Don't throw Hoyle on troubled waters." I saw Gamow lecture only once. He had been invited to Harvard by the astronomer Cecilia Payne Gaposhkin. (She was born in Britain but had married a Russian astronomer, Gaposhkin.) All during the lecture he kept referring to "*the* theory." Finally Gaposhkin got up the nerve to ask whose theory this was. Gamow looked at her as if the question bordered on the absurd and replied, "*Mine!*"

The application that Gamow made of the quantum theory in 1928 is relevant to the subject of this chapter – fusion. To be fair, the same work was done a little later independently by the physicists E. U. Condon and R. W. Gurney. To be even fairer, the same general idea was first discussed by Oppenheimer in connection with the theory of how electrons escape from the surface of metals. But it is buried so deeply in Oppenheimer's paper that it is easy to miss. The question that Gamow tackled was what the mechanism for alpha-particle decay is and how it accounts for the very strange properties of this decay. When an alpha particle emerges from the decaying nucleus it has a kinetic energy on the order of a few million electron volts. What is absolutely amazing is that the half-life of the various alpha-particle decays is an incredibly sensitive function of this energy. To take an example, plutonium-236 produces an alpha particle with an energy of 5.8 million electron volts. Its half-life is 2.9 years. On the other hand, plutonium-244 produces an alpha particle with a 4.6-million-electron-volt energy. But it has a half-life of 83 million years. Indeed, if one thinks about it, it is remarkable that alpha-particle decay happens at all. To explain the dilemma, I am going to analyze the reverse process – the fusion of an alpha particle with a nucleus. This poses the same problem and is closer to the application that interests us.

The alpha particle is, as we have learned, the nucleus of the helium atom. It has two protons and two neutrons. Each proton carries a positive electric charge, so the total charge is two in suitable units. The nucleus that we want the alpha particle to fuse with has, let us say, Z protons. For plutonium, Z would be 94. But like charges repel. So when the alpha particle comes close to the nucleus it feels these positive charges and is repelled. Figure 30 presents a diagram that shows how the electric energy that the alpha particle feels depends on the distance from the nucleus. You see that it falls off when the alpha is far away from the nucleus and reaches a maximum at the edge of the nucleus. The quantity E is the energy of the alpha. As drawn, this energy is lower than the maximum electric energy. If the alpha were a classical particle this would be the end of the story. It would not have enough energy to enter the nucleus over the energy barrier and would presumably go back to where it came from. But the alpha is not a classical particle. It is a quantum mechanical particle and, as such, is subject to Heisenberg's uncertainty

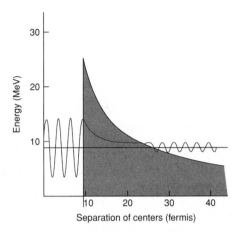

Figure 30. Tunneling model of alpha fusion. The alpha penetrates the barrier.

principles. One of these principles involves the energy measured in a process and the time it takes for this process to occur. It states that energy conservation can be violated if the process is short enough. The greater the violation, the shorter the allowed time for the process. In the case at hand, the alpha particle is allowed to tunnel *through* the barrier – a violation of the conservation of energy – if it does so rapidly enough. The theory allows you to calculate the probability of this tunneling as a function of the energy of the alpha.

If we look at the barrier in Figure 30 we see that the lower the energy of the incoming alpha, the more barrier it has to tunnel through. The more barrier it has to tunnel through, the less probable the tunneling. If we now look at the alpha-particle decay from the quantum point of view, the alpha particle rattles back and forth inside the nucleus. Periodically it hits the potential barrier. Each time, there is a probability that it can tunnel through. The smaller its energy the thicker the barrier it has to penetrate and the less probable is the tunneling. Thus isotopes that produce alpha decays with less energy are longer lived. Explanation of the very dramatic

relationship between energy and half-life requires a real calculation – the one that Gamow did.

The diagram in Figure 30 applies as well to the fusion of the alpha
particle with some nucleus. The barrier penetration issues are the
same. I want to begin the discussion of fusion that will eventually
lead us to the hydrogen bomb by beginning with a simple example –
hydrogen. It is a bit ironic – as we shall see – that this particular
fusion reaction is one that cannot be used in a so-called hydrogen
bomb, although the Sun exploits it to keep shining. We shall consider the fusion of two ordinary hydrogen atomic nuclei – protons –
to produce a nucleus of heavy hydrogen – the deuteron. You will
immediately have an objection. The two protons have together two
units of positive charge while the deuteron, which is one proton
and one neutron, has one unit of positive charge. What has happened to the other charge? In reality, this fusion reaction produces
not only a deuteron but also a positive electron and a neutrino. In
symbols:

$$P + P \rightarrow D + e^+ + v.$$

The positive electron – the "positron" – was the first anti-particle to
be discovered. The American physicist Carl Anderson observed it
in 1932. It has the same mass as the electron but the opposite charge.
The neutrino is a ghostly particle that is produced in beta decay. It
has no charge and a tiny mass. It hardly interacts with anything.

The first thing to remark about this reaction is that, in addition to
the particles, it produces energy. This is because the deuteron is less
massive than the two protons, which reflects the fact that the neutron and proton in the deuteron are bound together. It takes energy
to get them apart. The energy that is produced in this reaction –
about 1.4 million electron volts – is shared among the deuteron,

the positron, and the neutrino, with the neutrino getting the largest share. For the moment I am going to ignore the extra particles and focus on the fusion of the two protons. Since they have the same charge they repel each other electrically. This produces an energy barrier. Very roughly, how high is it? We want to make something the size of the deuteron, which turns out to be about one fermi. In case you have forgotten what a fermi is, it is 10^{-13} centimeter. When the two protons are placed this distance from each other the energy of the electric repulsion is about one million electron volts. Now we can imagine the following situation. We have an entire vat of protons and can heat them to any temperature we want. How hot is one million electron volts? The answer is about ten billion degrees. What does this mean? It means that if I have a vat full of protons whose average kinetic energy is about one million electron volts, then the temperature of the protons in the vat will be about ten billion degrees. Does this mean that to generate this fusion reaction we need a vat in which there are protons that have a temperature of ten billion degrees? Fortunately not, since the interior of the Sun is only fifteen million degrees. What have we forgotten? We have forgotten Gamow. The two protons can tunnel though the barrier when they have considerably less energy than the maximum. Moreover, we were talking about averages. In a vat of protons at a lower temperature there are always some with a kinetic energy greater than the average and these can tunnel. Putting these effects together, this fusion reaction takes place at temperatures on the order of the interior temperature of the Sun and other stars, which is fortunate because we need the energy. The proton-proton fusion is just the trigger for the sequence of fusion reactions in the Sun that produces the solar energy. I will come back to the rest of the sequence after I tell you a little of the history of the stellar energy question.

The first person to suggest that the source of stellar energy might be nuclear was Arthur Eddington. He was born in Kendal, England, in 1882, the son of a Quaker family. Eddington refused military service in the First World War and had no problem leading an eclipse expedition in 1919 that verified Einstein's theory of gravitation, even though Einstein was a "German." Eddington was one of the first people to master Einstein's general theory of relativity and gravitation and he wrote a very important monograph on the theory. He was once asked whether it was true that only three people in the world understood the theory. "Who is the third?" he replied. By the 1930s he claimed to have discovered a Theory of Everything, which was the cause of much ridicule among scientists. His reputation was diminished, so it is easy to forget that he was one the greatest astronomers and astrophysicists who ever lived. In 1920 Eddington gave a presidential address that he called "The Internal Constitution of the Stars" to a section of the British Association. In it he explained why a new source of energy was needed. He asked,

What is the source of the heat which the Sun and stars are continually squandering? The answer given is almost unanimous – that it is obtained from the gravitational energy converted as the star steadily contracts. But almost unanimously this answer is ignored in its practical consequences. Lord Kelvin [William Thomson] showed that this hypothesis, due to [Hermann von] Helmholtz, necessarily dates the birth of the Sun about 20,000,000 years ago; and he made strenuous efforts to induce geologists and biologists to accommodate their demands to this time-scale. I do not think they proved altogether tractable. But it is among his own colleagues, physicists and astronomers, that the most outrageous violations of this limit have prevailed. ... No one seems to have any hesitation, if it suits him, in carrying back the history of the earth long before the supposed date of the formation of the solar system; and in some cases at least this appears to be justified by experimental evidence which it is difficult to dispute. Lord Kelvin's date of the creation of the Sun is treated with no more respect than Archbishop Ussher's.[1]

James Ussher was a sixteenth- and seventeenth-century cleric who, by counting generations from Adam to Christ, concluded that the Earth had been created at 9 A.M. on October 3, 4004 B.C. In short, the evidence was that the Sun is billions of years old. Where does the energy come from?

Eddington made a proposal. "A star is drawing on some vast reservoir of energy by means unknown to us. This reservoir can scarcely be other than the sub-atomic energy which, it is known, exists abundantly in all matter; we sometimes dream that man will one day learn how to release it and use it for his service. The store is well-nigh inexhaustible, if only it can be tapped. There is sufficient in the Sun to maintain its output of heat for 15 billion years." Eddington died in 1944, so he lived long enough to learn about the discovery of fission, but not long enough to learn about the atomic bomb. He goes on,

Certain physical investigations in the past year [1920], which I hope we may hear about at this meeting, make it probable to my mind that some portion of this sub-atomic energy is actually being set free in the stars. F. W. Aston's experiments [he of the mass spectrometer] seem to leave no room for doubt that all the elements are constituted out of hydrogen atoms bound together with negative electrons. [Rutherford's idea that the neutron was a bound electron and proton is being invoked.] The nucleus of the helium atom, for example, consists of 4 hydrogen atoms bound with 2 electrons. But Aston has further shown conclusively that the mass of the helium atom is less than the sum of the

[1] See, for example, jet.efda.org/pages/content/news/2005/yop/augo5.html.

masses of the 4 hydrogen atoms which enter into it; and in this at any rate the chemists agree with him. There is a loss of mass in the synthesis amounting to about one part in 120. [This number is a little off. It is more like one part in 150. At the time, they did not have a really accurate mass of either the proton or the helium nucleus. But the idea is correct and very important.] . . . I will not dwell on his beautiful proof of this, as you will no doubt be able to hear it from himself. Now mass cannot be annihilated, and the deficit can only represent the mass of the electrical energy set free in the transmutation. [This is a somewhat odd way of saying that, according to Einstein, the energy liberated in any form is equal to the mass difference between the four protons and the helium nucleus multiplied by the square of the speed of light.] We can therefore at once calculate the quantity of energy liberated when helium is made out of hydrogen. If 5 percent of a star's mass consists initially of hydrogen atoms, which are gradually being combined to form more complex elements, the total heat liberated will more than suffice for our demands, and we need look no further for the source of a star's energy.

A little later he says, "If, indeed, the sub-atomic energy in the stars is being freely used to maintain their great furnace, it seems to bring a little nearer to fulfillment our dream of controlling this latent power for the well-being of the human race – or for its suicide."

In reading this kind of prescience I am again reminded of Leonardo da Vinci and his drawings of flying machines (see Figure 31). Leonardo's idea was certainly right but it was beyond the technology of the period. Eddington's idea is certainly right, but it was beyond the ability of the science of 1920 to implement. One needed the quantum theory. Indeed, the first attempt to apply the theory to stellar energy came in 1928, when Gamow presented the theory of barrier penetration. Gamow had a physics chum in Göttingen named Friedrich "Fritz" Houtermans. Houtermans was

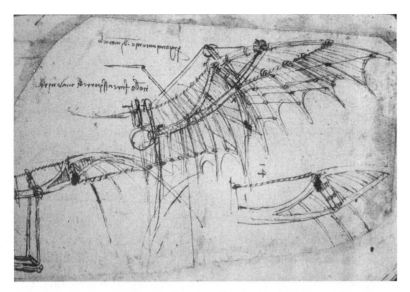

Figure 31. Da Vinci's flying machines. Art Resource, NY.

born in Danzig – now part of Poland – in 1903, making him a year older than Gamow. His father was a wealthy Dutch banker. Houtermans was so eccentric that if you put him in a novel and dealt only with the truth of his life, no one would believe you. In 1919, he was expelled from his Gymnasium in Vienna for reading the Communist Manifesto in the lobby of his school on May Day. His mother decided that he should be analyzed by Freud, who also expelled him after Houtermans confessed that he had been making up his dreams. He had enough discipline, and more than enough brilliance, to be admitted to the university in Göttingen, where he ultimately got his Ph.D. Later he got a job in the university in Berlin. By then he was married and was having too good a time to take Hitler's menace – even though he had Jewish ancestry – too seriously. His wife did, and they moved to England, where Houtermans got a job in industry. He disliked his job, and he disliked England, so he decided, against everyone's advice, to move to the Soviet Union. When his baggage arrived the police wanted to know why it contained seven editions of the Bible. This was 1935, and soon the repression of scientists began. Houtermans was arrested in 1937 and spent two and a half years in Soviet prisons where, at least in the beginning, he was tortured and put in solitary confinement. He kept his sanity by doing difficult mathematical problems in his head, but he lost all his teeth. In 1939, Hitler and Stalin made their pact, and the next year Houtermans was returned to Germany, only to be arrested by the Gestapo. That might have been the end of him, but von Laue got wind of his arrest and fished him out of jail. He also got Houtermans a job that he thought would keep him out of trouble.

There was an entrepreneur named Manfred von Ardenne who had become rather wealthy because of some of his inventions in radio and television. Ardenne had an estate near Berlin in which he had created a private laboratory sponsored by the German post office. When the war began he decided to shift his attention to nuclear energy. What his intentions were was never clear. He invented some form of the calutron and may have used it to separate uranium isotopes. The Russians were sufficiently interested that when they took Berlin in 1945, they shipped Ardenne, and his laboratory, to the Soviet Union. Houtermans was sort of house theorist to the laboratory in Berlin, but he was long gone before the Russians could nab him. Houtermans had proposed that element 94 – he did not know that it had been discovered and named plutonium by Seaborg – was fissile. He gave, in a long paper that was privately circulated, the Bohr argument that plutonium-239 was the fissile isotope. He was so frightened by the implications of this result that he tried, unsuccessfully, to warn colleagues in the United States. About the same time, 1940, Weizsäcker made the same discovery. He tried to patent it and he did turn it over to German Army Ordnance. All during the war, Heisenberg made the production of plutonium one of the reasons for building a reactor. None of the Germans had the remotest idea of plutonium's properties because they never made any. Remarkably, Houtermans survived the war, but not a lifetime of smoking cigarettes. He died of lung cancer in 1966 in Switzerland.

Houtermans was present when Gamow created his barrier penetration theory of alpha-particle decay. He recognized that if you turned the process around, it became a theory of alpha-particle fusion. Houtermans made a calculation of how probable a process fusion was and came to the conclusion that at any temperatures that were then available in terrestrial laboratories a fusion reaction was so rare as to be unobservable. But, in Göttingen that summer, there was a young British astronomer named Robert Atkinson. He had attended lectures of Eddington's and knew about the internal temperatures of stars and realized that they were hot enough to make fusion reactions significant. He and Houtermans also understood why this was a potential mechanism for the generation of stellar energy. This is a very essential point that will also apply to thermonuclear weapons. To understand it I refer to the diagram in Figure 32. It is a plot of the binding energy per particle across the range of nuclear elements – the "curve of binding energy." What we want to focus on is the fact that this binding energy increases – sometimes with a few jumps – until we come to iron, and then it decreases steadily. When we consider fission we move back up the curve from the heavy nuclei such as uranium to the more tightly bound lighter fission fragments. This is why energy is given off in fission. There is a mass loss due to the tighter binding. But look at the nuclei with masses less than iron. These are more tightly bound as we move up the mass scale. This means that if we fuse two light nuclei to make

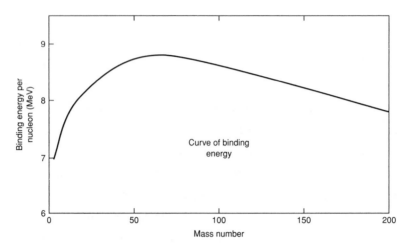

Figure 32. The curve of binding energy.

a heavier one, there will be a mass loss and, again, energy will be released. This is the key to everything, but the devil, as usual, is in the details.

Houtermans and Atkinson began trying to fill in the details with some help from Gamow. As it happened they had the right idea, but they were a few years too early. To account for solar energy they needed the neutrino, which Pauli did not invent until 1930. They also needed the positron which, as I mentioned, was not discovered until 1932, the same year as the discovery of the neutron. They also needed ideas from Fermi's theory of beta decay , which was not invented until 1934. As it was, they proposed some stellar fusion reactions that really don't work. But the paper that Atkinson and Houtermans wrote did have one important idea that endured. I have already mentioned that if we have a vat of particles at some temperature T, then, from a microscopic point of view, we can, in principle, measure the energies of the molecules in the vat. The average energy is proportional to T. But there is a whole distribution of energies that make up this average. Some molecules will have a smaller than average energy and some a larger than average energy. This distribution was first proposed in the nineteenth century by Maxwell – he of the demon. When you find the average energy you have to weigh the various energies with Maxwell's distribution. Likewise, when you find the probability of fusion of particles in a vat at a certain temperature, you have to use Maxwell's distribution to sum over the various energies of the fusing particles. As I discussed earlier, you find that fusions can take place at a lower temperature than simple energetics might suggest. Atkinson and Houtermans worked this out in general, and that result survived the rest of their paper.

It took a decade before the whole problem of stellar energy was really solved. The principal actors were Hans Bethe and Weizsäcker, with supporting roles played by Teller and Gamow and, especially, by Gamow's student Charles Critchfield, who worked in collaboration with Bethe. There was a wonderful collegiality at the time. There weren't many physicists. Stellar energy is a rich and wonderful subject, but I am writing about nuclear weapons and not astronomy, so I am going to limit myself to the aspects of it that are relevant to the matter at hand. Weizsäcker studied the consequence of starting the sequence of fusion reactions in the Sun with the fusion reaction we presented at the beginning of this chapter; namely proton plus proton goes into deuteron plus positron plus neutrino or, in symbols:

$$P + P \rightarrow D + e^{+}v.$$

Here he thought that there was a problem. Leaving aside gravitation, there are, as far as we know, three basic kinds of forces in nature. The strongest force is the one that holds protons and neutrons together. If it wasn't stronger than the next weakest force – electromagnetism – there would be no stable nuclei because the repulsive electric force among the protons would break the nucleus apart. The weakest force is the one that involves neutrinos. Any time that you see a neutrino in a reaction you can be sure that the reaction is very slow. Thus Weizsäcker, and others, thought that proton-proton fusion was too improbable to play an essential role in the generation of stellar energy. What Bethe and Critchfield did was calculate all of this in detail and show that in fact it explained solar energy just fine. Here is the sequence of reactions, the lifetimes associated with them, and the energy produced. There is more than one branch; I am only going to give the most probable one.

We begin with

$$P + P \rightarrow D + e^+ v.$$

Because this involves a neutrino it is very slow – indeed it takes about a billion years. This is why the Sun is so long-lived. The proton-proton fusion generates an energy of about 1.4 million electron volts. There are so many protons in the Sun – some 10^{57} – that this reaction, even though weak, is significant. The next step is for the deuteron to fuse with one of the myriad of protons. In words, this is deuteron plus proton goes into helium-3 plus a gamma ray; in symbols:

$$D + P \rightarrow {}^3He + \gamma.$$

Helium-3 is a light isotope of helium with two protons and a neutron in its nucleus. The gamma is a very energetic quantum of electromagnetic radiation. This reaction takes about a second and produces

5.49 million electron volts of energy. At this point there are three branches; I am only going to follow the most probable one, which occurs about 86 percent of the time. In this branch two helium-3 nuclei find each other and fuse into a helium-4 nucleus and two protons. In symbols:

$$^3\text{He} + {}^3\text{He} \rightarrow {}^4\text{He} + \text{P} + \text{P}.$$

This produces an energy of 12.86 million electron volts. So the combined energies in the cycle are about twenty million electron volts. It is a cycle because you begin with two protons and end with two protons. The time scale of this last fusion reaction is determined by two competing effects. On the one hand this is a fusion that involves the strong interaction, so it is almost instantaneous once it begins. But it requires two helium-3 nuclei and these are relatively rare. The time for this interaction in the Sun is about one million years. In their paper, Bethe and Critchfield, taking advantage of work done by Gamow and Teller, show that this cycle explains the observed solar energy production. I find it somehow satisfying that this cycle uses all three of nature's forces – the strong, the electromagnetic, and the weak. It also implicitly uses the gravitational force because that is what compresses the Sun, producing the temperature needed to start and sustain the fusion reactions.

All of this was done just before the war. When the war started, Teller went to Chicago and Bethe began working on radar. In the spring of 1942, Bethe received a phone call from Oppenheimer, who was at Berkeley. It was an open line but Oppenheimer managed to convey to Bethe that he was in charge of the theoretical development of a fission weapon and that there was going to be a summer session that he was inviting Bethe to join. Teller was also invited. In those pre-jet days one took the train. Bethe first stopped in Chicago to collect Teller. He also saw Fermi's reactor, which was under construction. He learned enough about it so that he was sure that it would work and that a nuclear weapon was a serious possibility. He and Teller shared a compartment on the ride from Chicago to California so that they could talk about classified matters. Teller explained to Bethe about plutonium and how the reactors were going to be used. Beginning in the fall of 1977, and lasting for about a year and a half, I did a series of interviews with Bethe that resulted in a *New Yorker* profile.[2] We spent a lot of time discussing Bethe's war-time activities, especially at Los Alamos, where he headed the theoretical division. He also told me about that train ride. He recalled, "Teller told me that the fission bomb was all well and good and, essentially, was now a sure thing. In reality, the work had hardly begun. Teller likes to jump to conclusions. He said that what we really should think about was the possibility of igniting deuterium by a fission weapon – the hydrogen bomb. Well the whole thing was far more difficult than we thought then."[3] Nonetheless, by the time they reached Berkeley they had at least the rough outlines of how a hydrogen bomb might work.

The first point that they understood was that they did not have to be bound by the reactions that take place in the Sun. In the Sun you are stuck with protons, so the first step involves a weak interaction that is extremely slow and can never lead to an explosion. On Earth, you can manufacture your starting materials. They came to the conclusion that the ideal fuel for a fusion weapon would be

[2] See *Hans Bethe, Prophet of Energy*, by Jeremy Bernstein, Basic Books, New York, 1979.
[3] Bernstein, op. cit., p. 73.

a mixture of deuterium and tritium. Deuterium we have discussed. Tritium is super-heavy hydrogen with two neutrons and a proton in its nucleus. It is unstable to beta decay with a half-life of about 12.3 years. Because it is unstable it does not exist naturally and must be manufactured in a reactor. For example, if a reactor uses heavy water – deuterium – as a moderator, there are enough neutrons around so that the deuterium can capture one of them, becoming tritium. Bethe and Teller also thought up a second mechanism for tritium production that involved lithium, in particular, the isotope lithium-6, with three neutrons and three protons in its nucleus. If there were neutrons around, then the reaction

$$^{6}\text{Li} + \text{N} \rightarrow \ ^{3}\text{H} + \ ^{4}\text{He}$$

would produce tritium. (Lithium-7 can also be used but it requires neutrons of a higher energy, which may be less abundant.) This meant that if you add lithium-6 to the deuterium-tritium mixture you can breed more tritium as you go along. The fusion reaction of interest in nuclear weapons is deuteron plus triton fusing into helium plus a neutron:

$$\text{D} + \text{T} \rightarrow \ ^{4}\text{He} + \text{N},$$

yielding an energy of 17.588 million electron volts, most of which goes to the kinetic energy of the neutron. This fusion involves only the strong interactions and hence is more or less instantaneous. Moreover, it has an ignition temperature of some tens of millions of degrees with an increased fusion probability as the temperature is raised to one hundred million degrees. They realized that this was the working temperature of a fission bomb and hence such a bomb could ignite a deuterium-tritium mixture. They also realized that there was a mass advantage in fusion over fission. As we have seen, each fission produces about 200 million electron volts of energy. By comparison this fusion reaction produces less than a tenth of the energy, but because it uses hydrogen isotopes and not uranium, it uses about one fiftieth of the mass, more than making up for the energy difference. Fifty grams of D plus T has about the mass of one gram of uranium. Furthermore there is no critical mass. You can use as much or as little of the stuff as you want. In short, it seemed perfect.

The idea of what came to be called the "classical super" was to take a container containing, for example, deuterium and tritium and possibly some lithium-6 and explode a fission bomb in its proximity. The problem was the distinction between "ignition" and "propagation." There is no question that if one simply exploded a fission bomb in such a container it would ignite some fusion reactions. But would the material in the container cool off before the energy from these fusions ignited others? You have the same problem when you try to get a log to burn. It is not hard to get some corner to burn, but then the fire goes out before the rest of the log catches. During the summer of 1942, Bethe and Teller engaged in a kind of tug-of-war as to whether one could get the "classical super" to explode. As Bethe put it to me,

About three quarters of our time that summer was occupied with thinking about the possibility of a hydrogen super-weapon. We encountered one difficulty after another, and came up with one solution after another – but the difficulties were clearly in the majority. My wife knew vaguely what we were talking about, and on a walk in the mountains in Yosemite National Park she asked me to consider carefully whether I really wanted to continue to work on this. Finally, I decided to do it. It was clear that the super bomb, especially, was a terrible thing. But the fission bomb had to be done, because the Germans were presumably doing it.

Bethe decided to continue with the fission bomb but not work on the hydrogen bomb until 1951, when its development seemed inevitable. But for Teller it was an obsession. It was the only thing he wanted to work on at Los Alamos. He would not join the rest of the laboratory in working on implosion, although he was asked to do so by Bethe, when the plutonium crisis arose. Oppenheimer, who was busy beyond human endurance, had to spend time every week with Teller, listening to his latest failed ideas for making the super. I have never really understood Teller's obsession. Why were fission bombs not enough for him? Was it the intellectual challenge? Was he worried that the Russians would get there first, although this was probably not a consideration during the war? Was he angry that Oppenheimer had made Bethe head of the theory division instead of him and he wanted to carve out a new domain? A combination? I don't know. I have asked myself what would have happened if, at Nagasaki, we had dropped a twenty-megaton hydrogen bomb instead of a twenty-kiloton fission bomb. The fission bomb killed 74,000 people and devastated everything within a radius of one mile from ground zero. A twenty-megaton bomb would have inflicted third-degree burns on everyone within a distance of twenty miles and would have devastated everything up to a distance of fourteen miles. Would the Japanese have surrendered sooner? Nagasaki was bombed on August 9th and the Emperor surrendered on the 14th. What would a hydrogen bomb have accomplished? In the years following the war, Oppenheimer often said that the problem with the hydrogen bomb was that the targets were too small.

It is not my intention to give anything like a technical history of the development of the hydrogen bomb if for no other reason than I think it is impossible. Enough of the technical information on the fission weapons has been unclassified – viz Serber's primer – that one has no problem accessing the relevant material. With the hydrogen bomb, almost everything is classified. If there exists anything like Serber's primer you would need a Q-clearance to read it. I only have access to the open sources. In any event, what I want to do is to give you a feeling of how the bomb works – not how to make one.

In 1946, there was a conference at Los Alamos at which Fermi gave a series of lectures on the progress, or lack thereof, of Teller's group, which was still working feverishly on the classical super, although the war was over. The lectures show Fermi's mastery of the physics and his ability to always get to the heart of the matter.[4] Interestingly, the work of Teller's group, and Fermi's analysis, was on deuterium-deuterium fusion and not deuterium-tritium fusion. The reason for this appears to have been economics. Tritium was thought to be too expensive to use in quantity. As I mentioned, it is manufactured in a reactor. A reactor that can produce a kilogram of tritium can, with the same number of neutrons, produce about seventy kilograms of plutonium. At the time, plutonium was scarce. The deuteron-deuteron reactions they considered produced tritium and a proton along with a relatively small four million electron volts of energy:

$$D + D \rightarrow {}^{3}H + P.$$

[4] I am grateful to Michael Goodman of Kings College for providing me with a copy of the notes on these lectures taken by P. B. Moon, who was a member of the British delegation to Los Alamos.

They noted that if they could get this reaction going, the tritium produced would fuse with the deuterium, adding to the energy production. The problem was getting the first reaction going. P. B. Moon's notes summarize the situation as it was in 1946 by stating that "So far all schemes for initiation of the super are rather vague." So it remained. Whether any scheme for the classical super would have worked is an open question. None was ever tested and, by 1951, when Stanislaw Ulam and Teller finally saw how to make a hydrogen bomb this was all academic.

While Teller and his group were knocking their heads together trying to make the classical super work, there was another activity going on that did not seem to attract much attention. This was the work of Klaus Fuchs and von Neumann. Klaus Fuchs, of whom we will hear much more in the next chapter, came to Los Alamos as part of the British delegation in August 1944. He had already been working on aspects of the bomb in England and soon proved himself to be an invaluable member of Bethe's theoretical division. He was known to have a photographic memory and to have involved himself in all aspects of the program. Almost no one was better informed. He was also a Russian spy – surely one of the most successful that has ever lived. By the fall of 1945, he had turned over to the Russians what amounted to a detailed blueprint of the gadget, whose test he had witnessed. The Russian physicists were ordered to duplicate the gadget, which they did. It was successfully tested in August 1949. Considering the effort the Russians put in and the ability of their scientists, it is certain that they would have gotten the bomb sooner or later. Fuchs probably saved them a couple of years. But, in 1946, Fuchs had turned his attention to the super. He and von Neumann patented their work and, as far as I can tell, the patent is still classified. However, in the spring of 1948, Fuchs turned it over to the Russians. Some of what he turned over has found its way back here by the back door, so to speak. In particular there is a diagram that has now been widely circulated.[5] Much of this diagram is of no interest because it is connected to the classical super. But a part of the diagram is of great interest because it is the first inkling of how to make a hydrogen bomb. After Fuchs was exposed as a spy, both Bethe and Oppenheimer belittled anything that Fuchs could have told the Russians about the hydrogen bomb. They even said that they hoped that the Russians would use Fuchs as a guide because that would lead them down a blind alley. I have often wondered if they actually knew what Fuchs turned over and, if they had seen this diagram, whether they still would have been quite so cavalier.

To see what Fuchs and von Neumann had in mind let us go back to the Sun. As we have seen, the Sun produces its energy by a cycle of fusion reactions. But these cannot get started until the central temperature of the Sun reaches a critical temperature. How does it do this? The Sun presumably began its life as a gaseous blob. But then gravitation began collecting the blob into a sphere and causing the sphere to compress. As it is compressed, the pressure at the center became greater and greater, which increased the temperature. When the threshold temperature was reached, the fusion reactions began and are still, fortunately, going on. The key then was

---

[5] See, for example, *The Brotherhood of the Bomb*, by Greg Herken, Henry Holt and Company, New York, 2002, the picture section after p. 210. Herken acknowledges Joseph Albright and Marcia Kunstel as his source. They were correspondents in Moscow, where they obtained the document. The diagram they obtained had Russian captions, which means that it is a translation of what Fuchs transmitted. I am very grateful to Carey Sublette for providing me with a version of the diagram in which the captions have been retranslated back into English. I am also grateful to him for explaining to me the significance of the diagram.

compression. I have no idea if Fuchs and von Neumann were thinking of the Sun when they decided that the secret of igniting fusion
reactions in a bomb was compression. You would not use the explosion of the primary to heat the fusible elements directly by agitat-
ing their molecules; rather you would use the energy of the primary
to compress something. To see how they proposed to do this I will
describe very schematically the diagram. There is a large container.
Inside it there is a gun barrel. Because the plutonium bombs at the
time were still fairly primitive, they were not suited to what Fuchs
and von Neumann wanted to do, so their diagram employs a gun-
assembly device. A modern hydrogen bomb would never use this.

Down the barrel there is a container, which was called the "spark
plug," in which a mixture of liquid deuterium and tritium is held.
When the bomb goes off the material from the bomb bangs into the
container and this gives it its initial compression. There is nothing
especially imaginative about this, but then there is a second compression that is supplied by the radiation from the bomb. This is the
true novelty. Radiation comprises quanta that carry both energy and
momentum. When radiation strikes a surface and the quanta bounce
off, a pressure is produced. That happens here, but this direct pressure is much too small to cause any real compression. But what it
does do is to knock all the electrons off the deuterium and tritium
atoms, as well as those of the material of the gun barrel. In the example used by Fuchs and von Neumann, this material is taken to be
beryllium oxide. Beryllium has four electrons and oxygen has eight,
so the number of particles an ionized molecule of beryllium oxide
produces is fourteen, including the two nuclei. On the other hand,
deuterium and tritium each produce, when ionized, a single electron, so this mixture produces four particles per ionization, including the nuclei. Therefore the ratio of particles is 14/4, or a little more

than three. Why does this matter? In this arrangement the temperatures of the beryllium oxide and the hydrogen isotopes remain the same. But in this situation the laws of thermodynamics tell us that the ratio of the pressures is equal to the ratio of the number of particles – about three in this example. Thus the deuterium-tritium mixture is subject to a pressure due to this ionization and this is what causes the second and definitive compression. Fuchs and von Neumann were confident that this would ignite the fusion of the deuterium and tritium. Teller was equally adamant that it would not. He gave some specious arguments as to why it would not work. I suspect his vehemence was a product of the fact that he had not thought of it.

By 1951, he had changed his mind. He designed the thermonuclear part of a test that became known as Greenhouse-George. As far as I can tell, the compression part was identical to what is in the Fuchs–von Neumann patent. It was a curious test because, by then, Ulam and Teller had invented a much better compression scheme. Oppenheimer objected to the George test on the grounds that it was pointless because the Ulam-Teller scheme, which was already known, would replace it. This was one of the arguments that was used in his 1954 trial to show that he had been dragging his feet on the hydrogen bomb. But he was right. The test, on May 8, was fired from a tower on Eberiru Island in the Enewatak Atoll in the Pacific (see Figure 33). It produced a 225-kiloton explosion – the largest ever up to that time. It also produced a good deal of fallout that, because of the winds, did not fall back on the island. But it was collected by the Russians, who were able to learn that it was a fusion test. They knew, if for no other reason than that Fuchs had given them a copy of the notes of Fermi's lecture about the virtues of the deuterium-tritium mixture. As I mentioned, this fusion also

Figure 33. The mushroom cloud from Greenhouse George (May 8, 1951), the first bomb that used fusion. Photo courtesy of National Nuclear Security Administration/Nevada Site Office.

produces energetic neutrons. It is virtually a neutron factory. These neutrons can be absorbed by some of the remaining plutonium. They can cause additional fission and thus this configuration acts as a "booster" for fission bombs. This means that less plutonium is required and the bombs are lighter and suitable to be put in missiles. Both tests that I saw in Nevada were of boosted weapons. I am sure that the clicking pump noises I heard had to do with the insertion of the deuterium-tritium mixture into the interior of the hollow pit. But the fusion neutrons can also be absorbed by the plutonium and produce transuranic elements. Both Fermium and Einsteinium were first discovered in the detritus from later hydrogen bomb tests. But any unusual distribution of transuranics in the fallout would mean that a high neutron density had been created in the explosion, a sure sign of fusion. Whether the Russians actually used the Fuchs–von Neumann patent to put two and two together, I am not sure. Material gathered from espionage was very tightly held and only a minuscule number of physicists were allowed to see it. I do not think that Andrei Sakharov, who was largely responsible for inventing the Russian hydrogen bomb, was in that group, but he was certainly in contact with people who were. Recent reviews of this history by Russian scientists who were involved give the impression that what Fuchs gave them was much more valuable than had originally been thought.

As I have suggested, I do not think it is possible to make a real historical accounting of the steps that led to the final hydrogen bomb design – the Ulam-Teller or Teller-Ulam design, depending on whose account you believe. The principals are dead; Teller died in 2003 and Ulam in 1984. Documents that might shed some light on this are classified. One has to settle for various people's conflicting accounts. Teller, in a way, had the last word in his memoir, published not long before his death.[6] His position was that Ulam had

---

[6] *Memoirs; A Twentieth Century Journey in Science and Politics*, by Edward Teller, Perseus, New York, 2002.

little to do with it and that he, Teller, deserves the credit. In his memoir he is still angry with Oppenheimer for making Bethe head of the theoretical division at Los Alamos instead of him. Ulam is, I think, a different case. He certainly had his share of ego. I once heard him give a lecture called "Unsolved Problems in Mathematics and Their Solutions." He was really a pure mathematician at heart. He was born in 1909 in what was then Lemberg, Poland. He began a serious self-study of mathematics when he was fourteen and got his degree in mathematics in 1933. His reputation as a mathematician was such that von Neumann invited him to visit the Institute for Advanced Study in 1935. It was von Neumann who arranged for Ulam to come to Los Alamos. After the Institute, he developed a relationship with Harvard that involved commuting back and forth to Poland. In 1939, he left Poland for the United States with his younger brother Adam, who eventually became a professor of government at Harvard. The rest of their family was killed in the holocaust. In 1943, Ulam became an American citizen and went to Los Alamos that year. He was, in a way, an unusual mathematician. Physicists often complain that when you ask a mathematician about a problem, they show you that it is equivalent to another problem that they also cannot solve. Ulam solved problems. Among other things, he invented a method of carrying out approximately very complex numerical calculations, which was called the Monte Carlo method. This was very valuable because of the extremely limited computing power available at the laboratory. After the war, he stayed on. I think that he had come to love the desert southwest. In 1950, President Truman ordered a crash program to make the hydrogen bomb and Ulam, with another Los Alamos mathematician named Cornelius Everett, made a serious attempt to see if the "classical super" could possibly work. They decided that it could not

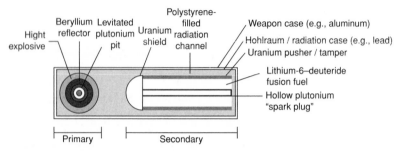

Figure 34. A rendering of an Ulam-Teller–designed hydrogen bomb.

with any amount of tritium that made sense, which drove Teller into a kind of frenzy. I think that Ulam kind of stumbled into making the first steps toward the design that did work.

Ulam proposed a two-stage device in which the primary atomic bomb stage was to be used to compress the container of the fusible elements. His first idea was to use the massive material from the bomb explosion to make this compression. It was Teller who then suggested that they use the radiation instead. I think that he was aware of the Fuchs–von Neumann proposal and understood its limitations. In their proposal, as I have mentioned, the heavy material of the container and its contents – the deuterium-tritium mixture – are kept at the same temperature. The laws of thermodynamics then limit the ratio of the pressures to the ratio of the ionized particles. This is the best you can do and the compression is therefore limited. Figure 34 shows a schematic diagram of the Ulam-Teller configuration – one that I got from the web.

Notice the salient features. There is a case for the whole device made out of aluminum. On the left of the interior is an implosion plutonium bomb. On the right is a cylindrical case made of uranium, a detail that is very important. Inside the case is another hollow tube called the "spark plug." Unlike the Fuchs–von Neumann spark plug, this one does not hold the fusible elements. The tube is hollow and made of plutonium. It looks, for good reason, like a cylindrical implosion bomb. Outside of it, in solid form, at least in modern weapons, is the deuterium–lithium-6 mixture. The sequence of events is the following: after the explosion of the bomb the X-rays fill the space above the capsule holding the spark plug and the lithium-deuteride. The uranium casing of the capsule is at a very high temperature. But the interior is blocked off from the X-rays so that it is not immediately heated. The casing now expands violently like the exhaust from a rocket. It exerts an enormous pressure on the plutonium, compressing it and making it supercritical. Fission begins, and this raises the temperature of the lithium-deuteride, igniting it. The fusion reaction produces energy, but also neutrons. These can boost the fission reaction. In hydrogen bombs at least 50 percent of the yield is produced by fission.

This arrangement may seem a little baroque, but it was immediately understood that it would work. Oppenheimer called it "technically sweet." It took until October 1952 before the design was tested. Meanwhile, Teller had left Los Alamos and Carson Mark was in charge of the development. The device that was first tested differed from the one that I just described in that liquid deuterium was used instead of lithium-deuteride. The test took place on Eluglab Island in the Enewetak Atoll, as part of the so-called Ivy series. This one was Ivy Mike. It took place on the 31st of October and produced a 10.4-megaton explosion, which eliminated the island. Eight megatons were from fission and the rest from fusion. The whole device weighed eighty-two tons. Incidentally, this explosion produced the element fermium, which has an atomic number of 100. The isotope of fermium produced had an atomic weight of 255. This means that to produce it, a very large number of neutrons had to be absorbed – seventeen, if you start from uranium-238. But this amount of neutron absorption is possible only if you have a very high density of material. Thus anyone who analyzed the debris and found fermium – which we kept secret – would know that compression was responsible for the fusion process.

The first "dry fuel" – that is, lithium hydride – bomb was tested on March 1, 1954, at the Bikini atoll on the Marshall Islands. It yielded fifteen megatons, which was the largest explosion that we ever produced. (The Russians exceeded it. They had a fifty-megaton explosion.) The fifteen megatons were a mistake. It was supposed to be half that or less. The designers did not realize that the lithium-7, which had been mixed with lithium-6, also produced tritium because the neutrons were so energetic. The fallout poisoned the crew of the *Daigo Fukuryü Maru*, a Japanese fishing boat. Given the history, the international outcry was certainly instrumental in the successful imposition of the aboveground test ban.

Finally, I want to discuss the question of whether the hydrogen bomb should ever have been built in the first place. Let us recall the events that led up to President Truman's decision. In August 1949, the Russians successfully tested their first fission bomb. On January 27, 1950, Fuchs confessed to his espionage. In October 1949, there was a four-day meeting of the General Advisory Committee of the Atomic Energy Commission. This committee consisted of the highest-level people in the weapons program. Rabi and Fermi were members, as was Conant. Oppenheimer was chairman. The committee decided for various reasons that there should not be a crash program to build the hydrogen bomb and, above all, *not* one that was publicly announced. Some members of the committee thought that the hydrogen bomb was not a weapon of war but a method of genocide. Rabi and Fermi thought there should be a conference with the

Russians to seek an agreement not to build it. If that failed, nothing would have been lost because, at the time, no one had a clear idea of how to build one anyway. But, after Fuchs's confession, the pressure on Truman to do something was irresistible, and four days later he publicly announced a crash program to build the hydrogen bomb. The predictable happened – a hydrogen bomb race. The Russians exploded theirs in 1952, the British in 1955, the Chinese in 1967, and the French in 1968. Probably every country that has atomic weapons is engaged in building the hydrogen bomb. There is some sad irony in all of this. In the same 1952 Ivy series there was a test called Ivy King. This was a pure fission bomb – no boosting – that had been designed by Theodore Taylor of Los Alamos. It produced a 500-kiloton yield. Now here is the irony. In recent years the Russian and American nuclear strategists have concluded that megaton bombs are unnecessary. The arsenals have been cut back to bombs of several hundred kilotons. Put another way, if the hydrogen bomb had never been built and pure fission bombs had continued to be developed, then the Russian and American nuclear arsenals would look about the same as they do now.[7]

---

[7] I am grateful to Freeman Dyson for calling my attention to this possibility and for discussions. Ted Taylor was our "boss" on the Orion project, which was an abortive attempt to use nuclear bombs as propellants for a very large spaceship. The test ban put an end to the idea.

11. Spies

Figure 35. Klaus Fuchs. Courtesy of the National Archives.

A S FAR AS I KNOW, DURING THE WAR THERE WAS NO
German atomic spy. I think the reason is that, on any official level, the Germans did not take seriously the possibility that the
Allies might be working on an atomic bomb. There were three pro-
grams in Germany, with some overlap, that were working on nuclear
energy. There was what might be called the "official" program. It
had started in September 1939 and was under the auspices of the
Heereswaffenamt – German Army Ordnance – that was located in
Berlin. They had the power of the draft and, among others, they
drafted Heisenberg and Weizsäcker. At its zenith it involved some
forty physicists and chemists – the *Uranverein* or the "Uranium
Club" – spread over nine different sites. Compared to the Manhattan Project it was minuscule. In the beginning, there was a mandate
to look into the prospects for making a nuclear weapon, but quite
soon Heisenberg, who was its intellectual leader, focused on designing a reactor. Although he was one of the greatest theoretical physicists of the twentieth century, he lacked practical engineering experience. In short, he was no Fermi. Whereas the papers he wrote on
these designs are very impressive, he generally chose configurations
that were relatively easy to calculate rather than more inelegant

configurations that worked better. In February 1942, with the war going badly, the army pulled out because there seemed to be no prospect of making a weapon in the foreseeable future. The project was taken over by the Reich Research Council, and Heisenberg became the de facto director. In February 1942, the group published a 200-page report on what they had accomplished. There is scarcely a mention of nuclear weapons.

The second project was von Ardenne's, which took place in a private laboratory, on his Berlin estate. There was some disagreement on the connections this project had with the official one. Heisenberg and Weizsäcker claimed that there was essentially none, but Ardenne kept a visitor's book containing signatures of the visitors. The book showed visits from Hahn, Heisenberg, and Weizsäcker. Ardenne claimed that, in the late fall of 1941, he had successive visits from Heisenberg and Hahn, who told him independently that the critical mass of uranium-235 was about ten kilograms. He also reported that, on a later visit, Weizsäcker informed him that Heisenberg had decided that a uranium bomb was impossible. One has no idea what to make of any of this. Nor is it clear what Ardenne's laboratory was designed to do. It is hard to imagine that separating uranium isotopes and circulating memoranda on the fissibility of plutonium were being done for peaceful purposes, which is what Ardenne tried to claim in his memoirs. Houtermans seems to have had an inkling that there was an Allied project, but he certainly did not have the means, let alone the inclination, to organize any espionage.

Until recently, those were the only two German nuclear energy projects anyone knew about. But in 2005, the German historian Rainer Karlsch published a book in which he presented evidence for a third.[1] It seems that after the army stopped supporting the official project, it did continue to support a secret project in Berlin. It was led by the physicist Kurt Diebner, who was a member of the Nazi Party and an expert on explosives. This project, which later reported directly to Heinrich Himmler, was certainly trying to develop a nuclear weapon. How far they got is a matter of controversy. Interestingly, Diebner and several other members of the *Uranverein* got the opportunity, starting on July 3, 1945, to spend six months together as guests of the British in a manor house near Cambridge named Farm Hall. This came about through the following sequence of events.

In the fall of 1943, General George Marshall, the U.S. army chief of staff, decided that he was not satisfied with the desultory intelligence reports he had been getting about the German nuclear program. With the German army in retreat, he proposed a mission that would follow the army closely and would concentrate on learning about the nuclear program. He gave General Groves the task of creating the mission. Someone named it "ALSOS." I am not aware that Groves was a Greek scholar. When he learned that αλσóc was Greek for "grove" he was not amused. The military side of the mission was commanded by a Lieutenant Colonel Boris Pash. Pash had a touch of derring-do. His jeep was the second to enter liberated Paris and he occasionally got ahead of the regular troops. On the scientific side, the mission was headed by the Dutch-born American physicist Samuel Goudsmit. Goudsmit was an excellent choice. He spoke several European languages and had known people like Heisenberg from before the war. He was also highly motivated. Both

[1] *Hitler's Bombe*, by Rainer Karlsch, DVA, Munich, 2005.

of his parents had been murdered in Auschwitz. He often said that another of his virtues was that he did not know very much about our project. If he had been captured, he reasoned, he had nothing much to tell. But this was also, as it turned out, something of a disadvantage. Goudsmit did not know anything about plutonium, so it did not occur to him to ask any of the German physicists such as Heisenberg and Weizsäcker about it. When he wrote his book he accused them of incompetence for not having seen this alternative, which they perfectly well had. This led to an angry exchange with Heisenberg that did not do either of them much good.

The mission first went to Italy, where they did not learn much, and then France. In Paris they visited Joliot's cyclotron, where they discovered that they had been preceded by a long list of Germans including Bothe and Diebner. What any of the Germans got from Joliot is not clear, but probably very little that had any relevance to nuclear weapons. Diebner wanted the cyclotron to be moved to Germany but Joliot, who was in the Resistance, did not. As a compromise, a German team was left behind to run it. As it happened, its head, Wolfgang Genter, was an anti-Nazi. As far as I know, Joliot never separated any uranium isotopes, nor did he make plutonium. The mission then went on into Germany. By December they were in Strasbourg (Strassburg in German). When the Germans occupied Strasbourg they Aryanized the university. Weizsäcker had gotten a job there. He was nowhere to be found, but he had left behind a treasure trove of documents. One of them was the letterhead of a letter from Heisenberg with his street address and telephone number. Goudsmit entertained the idea of going to Switzerland and calling him.

When Goudsmit returned to the United States for a Christmas leave he was sure, from what he had found in Strasbourg and elsewhere, that the Germans did not have a significant nuclear weapons program. General Groves was informed. He must have wondered who to tell this to. The motivation of many people at Los Alamos was that they were in a desperate race with the Germans. What would the reaction be if they knew that there was no race? For example, take Joseph Rotblat. He was born in Warsaw in 1908. He worked as an electrician and studied physics at night, getting a doctorate from Warsaw University in 1938. In 1939, he joined Chadwick in Liverpool. After fission was discovered – and Poland was invaded – he urged Chadwick to begin thinking about nuclear weapons because the Germans were certain, he thought, to be doing so. When Chadwick went to Los Alamos, so did Rotblat, even staying in Chadwick's lodgings. As it happened, Chadwick was one of the few Englishmen Groves, who was an Anglophobe, had any use for. When Groves came to Los Alamos that December, he told Chadwick what Goudsmit had found. Chadwick told Rotblat because he knew that the only reason that Rotblat had agreed to work on something that he found morally reprehensible was because he found the Nazis to be even more reprehensible. Rotblat promptly left Los Alamos. One of the conditions of his leaving was that he was not allowed to tell anyone the real reason.

After his Christmas vacation Goudsmit returned to Germany to finish his mission. By this time the Allies had moved well into Germany and as many important German scientists as the ALSOS people could lay their hands on were being rounded up and interrogated. It was decided to pick out a small group – ten in the event – and send them to Britain to be interned. Goudsmit selected them. It was an eclectic choice considering both whom it included and whom it left out. Among the latter was Bothe. He was a patriotic German but in no way a Nazi. His work suffered greatly during the war because of that. He refused to tell Goudsmit anything before Germany surrendered, but after the surrender, he told everything he knew. Goudsmit probably decided that there was nothing more to be learned from him. Among the selected were Heisenberg, Weizsäcker, and Diebner, as well as Hahn. Von Laue was selected certainly in part to protect him. His anti-Nazi activities were well known and there were still some overfanatical Nazis, known as werewolves, who might want to do him harm. Goudsmit also selected a young and very brilliant physical chemist named Paul Harteck. It was Harteck's 1939 letter to German Army Ordnance about the prospects of nuclear weapons that had gotten the whole project started. Harteck was entirely apolitical, but he needed money for his research projects and he thought, correctly, as it turned out, that this would be a good way of getting money from the government. During the war he invented a new kind of centrifuge for separating isotopes and had several ingenious ideas for making a reactor, but he was rather low on the totem pole in the project and could never get enough uranium to try them out. He ended his career in the United States at the Rennselaer Polytechnic Institute.

On July 3, this rather unlikely group was flown from Belgium to England in a Dakota and taken to Farm Hall. Some of them had been to Cambridge and recognized the surroundings. What none of them knew was that the manor house had been wired so that all of their conversations were overheard. The concealment of the wiring was such that when the new owner bought it after the war – a Cambridge don – he had no idea of its history. When I visited, he explained that he had had some of the floor boards taken up for renovations and was surprised to find a network of wires underneath. He later learned why they were there. He also told me that some of the detainees had come back from time to time for sentimental visits.

During their 1945 stay, their conversations were recorded on shellacked discs that were then reshellacked. It appears as if none of the original discs were preserved. There was a group of translators and transcribers who took the conversations off the discs and translated them. Periodically, a pared-down version of the transcripts was sent to people like General Groves. I have a photocopy of the version he read with his annotations. It is clear what interested him. Any mention of the Soviet Union by one of the detainees the general would underline with many exclamation points. On one of his visits to Los Alamos he said to Chadwick, in Rotblat's presence, that the real reason for carrying on with the bomb was to contain the Soviet Union.

One of the highlights of the transcripts is the reaction of the detainees to the news of Hiroshima. At first they refused to believe it. Weizsäcker said, "I don't think that it has anything to do with uranium."[2]

To which Hahn added, "It must have been a comparatively small atomic bomb – a hand one."

Heisenberg then entered the dialogue with, "I am willing to believe that it is a high pressure bomb and I don't believe that it has anything to do with uranium but that it is a chemical reaction where they have enormously increased the whole explosion."

From this I think it is quite clear why the Germans never tried to find out, during the war, if we had a nuclear program. They simply didn't believe it. The Farm Hall detainees, at least most of them, immediately began to justify themselves. Weizsäcker set the tone.

"History will record that the Americans and the English made the bomb, and that at the same time the Germans under the Hitler

2 *Hitler's Uranium Club*, with annotations by Jeremy Bernstein, Copernicus Press, New York, 2001, p. 117.

regime produced a workable engine [reactor]. In other words, the peaceful development of the uranium engine was made in Germany under the Hitler regime, whereas the Americans and the English developed this ghastly weapon of war."[3]

This from the man who on July 17, 1940, presented to German Army Ordnance a proposal to use plutonium as a nuclear explosive. Incidentally, the Germans never got an "engine" to work.

In contrast, our obsession with the Germans was there from the very beginning. The last paragraph of Einstein's 1939 letter to Roosevelt reads, "I understand that Germany has actually stopped the sale of uranium from the Czechoslovakian mines which she has taken over. That she should have taken such early action might perhaps be understood on the ground that the son of the German Under-Secretary of State, von Weizsäcker, is attached to the Kaiser-Wilhelm-Institut [sic] in Berlin where some of the American work on uranium is now being repeated."

One wonders where Szilard, who wrote the letter, got his information. It is true that C. F. von Weizsäcker was in Berlin at the Kaiser Wilhelm Institute. It is also true that his father, Ernst von Weizsäcker, who was later tried and convicted at Nuremberg – he was sentenced to seven years but was released in 1950 as a part of a general amnesty – was in rank in the foreign office just below the foreign minister Joachim von Ribbentrop. It is also apparently true that the younger Weizsäcker spoke to his father in a general way about the nuclear project. And it is true that his father approved the invasion of Czechoslovakia. I do not know if he was responsible for stopping the sale of uranium. What is off the mark is the notion that the Germans were primarily engaged in repeating American

---

[3] Bernstein, 2001, p. 138.

work on uranium. Because of Szilard's efforts, none of that work, after a certain point, was known to the Germans. They got started on trying to make a reactor two years before Fermi did. However, by 1939 there was a British agent in place, an Austrian named Paul Rosbaud.

Rosbaud was a love child born in 1896 in Graz who never knew his father, an organist named Josef Hennisser.[4] Hennisser, on his occasional visits to Graz, managed also to father Rosbaud's siblings. Rosbaud was brought up largely by a neighbor, a tram conductor named Johann Strajner. When the Nazis came into power, Rosbaud and his brother Hans needed a document that showed that they were Aryans through and through. Strajner, whose ancestry was impeccable, agreed to claim that he was their father. Their mother, a certified Aryan, had in the meantime died. As a nice touch, Strajner regularly put flowers on her tombstone. Rosbaud served in the army in the First World War and near the end of it was taken prisoner by the British, whom he came to like. He studied chemistry and, in 1926, took his Ph.D. He decided that he was more interested in the publication of scientific papers than in writing them. He became a kind of scout for a weekly publication on metallurgy named *Metallwirtschaft*. His job was to go around to universities and the like and solicit articles. The owner of the journal was a crypto-Nazi and after Hitler came to power revealed himself. Rosbaud quit and got a job at the prestigious publishing firm Springer, again as a scientific advisor. By that time he knew everyone who was anyone in German science.

---

[4] The standard biography of Rosbaud is *The Griffin: The Greatest Untold Espionage Story of World War II*, by Arnold Kramish, Houghton Mifflin, New York, 1986. It is a pity that Kramish never brought the book up to date with the more recent information on the German program. His discussion of plutonium, for example, is largely incorrect.

Hahn told him privately about the discovery of fission. Rosbaud arranged for the Hahn-Strassmann article to be published on an urgent basis in the Springer journal *Naturwissenschaften*, which meant re-editing an issue that was about to go to press. Rosbaud did not do this only for scientific reasons. He immediately understood the implications of the discovery and wanted people outside Germany to see the dangers. On March 10, 1939, he had lunch in Berlin with John Cockroft and told him everything the Germans were doing in fission. This is what started the British program, although it was the work of Peierls and Frisch that kept it going. I do not know if Cockroft communicated the details of this lunch to anyone in the United States. In 1938, Rosbaud was invited to England, where he had already sent his Jewish wife and daughter. He declined, feeling that he could be useful to the British in Germany. He was one of the people who helped Meitner to escape. Rosbaud became an agent for MI6 with the code name "Griffin." He transferred invaluable material mainly through Norwegian contacts. When, in 1942, the army stopped sponsoring the uranium project, Rosbaud knew, and he also knew of Heisenberg's meeting in 1942 with Hitler's newly appointed armaments minister, Albert Speer. He knew that Heisenberg had asked Speer for what Speer thought was a ludicrously small amount of money for a serious military program. He was used to projects like the building of rockets that involved thousands of people. All of this convinced Rosbaud that the Germans did not have a significant nuclear weapons program, which he conveyed to the British. They no doubt told Groves, but Groves had an assignment to build a bomb, and that was what he was going to do.

When I was interviewing Bethe I asked him if during the war the Los Alamos people had learned anything about the German program. He gave me the one example he could remember. It involved Bohr, and it is a perfect illustration of the limitations of this kind of intelligence. Until the fall of 1943, Bohr remained in Copenhagen. On January 25, 1943, he received via the underground a message from Chadwick. In it he invited Bohr to England to work on a "particular problem" that he thought would be of interest to Bohr. He mentioned some possible collaborators from which Bohr understood what the problem was – nuclear weapons. In February, Bohr replied that he still thought using the "discoveries of atomic physics" for this purpose improbable. In view of that, he felt that he was obligated to stay in Denmark to try to help out there. But sometime late that summer Bohr had a change of mind and told Chadwick in a communication that he was now convinced that nuclear weapons were a real possibility. What changed his mind was the second of two visits that the German physicist Hans Jensen made to Copenhagen. This one was in the summer of 1943. Jensen was not a member of the *Uranverein* – his politics were suspect – but he was doing related work on extracting heavy water and was in touch with the work of the group. It appears as if he got notes of a lecture that Heisenberg had given in May and had come to the conclusion that Heisenberg had presented the design of a nuclear weapon. This is what he told Bohr. A drawing was made by someone. It might have been Bohr, on the basis of what he heard, but I think that it was most likely Jensen. In any event, in September 1943, Bohr got word that he was going to be arrested by the Gestapo. He and his family were taken by boat to Sweden and then Bohr was flown to England, where he was extensively debriefed by people like Chadwick. He learned for the first time what the Allied program was like. I have never been able to learn if he showed Chadwick, or anyone else in England, the drawing.

On November 29, Bohr and his son Aage, also a physicist and later a Nobel Prize winner, sailed on the *Aquitania* for New York, arriving a week later. He then went to Washington, where he met

Groves, among others. On December 27th he went by train with Aage to Chicago, where he again met Groves, who accompanied them on the two-day trip to New Mexico. There is no doubt that he showed the drawing to Groves. Indeed, the general was so alarmed that he insisted that Oppenheimer stop everything and hold a meeting to analyze it. The meeting took place on December 31st and the participants were several of the senior physicists at Los Alamos. Among them, in addition to the Bohrs, were Bethe, Serber, and Teller. It was Serber who told me about the meeting, to which he arrived a little late. Oppenheimer handed him the drawing, saying something about Heisenberg, but offering no other explanation. Serber looked at it and immediately realized that it was a design for a reactor. Indeed, as he told me, the design looked a little "silly." This was because it was a Heisenberg reactor design in which he was still striving for elegance rather than practicality. The uranium was stacked in neat layers with the heavy-water moderator running through them. As I have noted in connection with the Fermi reactor, the most efficient design is to have the uranium in separated lumps or rods. Heisenberg finally adopted this design near the end of the war, and he might have gotten it to work if ALSOS had not stopped the project. When Bethe looked at the drawing he thought that the Germans had gone crazy and wanted to drop a reactor on London. He and Teller made an analysis and demonstrated that neither this, nor any other reactor, could blow up like a bomb. Their paper was sent to Groves and the matter put to rest. What struck me about all of this is that none of them thought of the real reason why the Germans wanted, at least in the beginning, to build a reactor – to make plutonium. Heisenberg repeated this on several occasions during the war. One can only wonder what the reaction of the people at Los Alamos, to say nothing of Groves, would have been if they had known this.

While the Germans were indifferent to the Americans and the Americans were obsessed by the Germans, the Russians were obsessed by both. Their attempts to learn about the American nuclear project began well before Los Alamos. In 1942, one of the operatives in the Soviet consul in San Francisco named Pytor Ivanov asked a British-born petroleum engineer named George Eltenton to find a way to contact Oppenheimer in order to learn about the Berkeley atomic bomb program. How Ivanov knew there was such a program, I do not know. Eltenton, in turn, contacted another member of the Communist Party named Haakon Chevalier. Chevalier was a professor of French literature at Berkeley and a very close friend of Oppenheimer's. At a small dinner at the Oppenheimers' Chevalier took Oppenheimer aside and conveyed Eltenton's request. Oppenheimer said in no uncertain terms that to do such a thing was wrong, and because we and the Russians were allies any information like this should be given to them, if at all, through official channels. Chevalier's intentions were, at least to me, never very clear. Was he conspiring with the Russians or was he, as he claimed, merely alerting Oppenheimer to something that he should know about? Oppenheimer then did something that was extremely foolish and later in his security hearings did him great damage. He told the security people about the request, but made up some story that concealed Chevalier's identity. Over the next few years, when confronted he invented new stories until General Groves finally made him tell the truth. Chevalier was interrogated and more or less forced to leave the country. When the full story emerged at the hearings, even Oppenheimer's friends were appalled.

At Los Alamos, as far as anyone knows, there were three, and only three, Soviet spies. In order of increasing importance they were David Greenglass, Theodore Hall, and Klaus Fuchs. I will begin with Greenglass. Greenglass was born in 1922 in New York. He had an early interest in communism, which he shared with his wife, his older sister Ethel Greenglass Rosenberg, and Julius Rosenberg, her husband. Greenglass and his wife joined the Young Communist League early in 1943, and soon afterward he was inducted into the army. He was sent to Jackson, Mississippi, where, as a skilled machinist, he was promoted to sergeant. He was first stationed at Oak Ridge and then sent to Los Alamos. According to the testimony he gave at his trial in 1950 – Fuchs's confession in February 1950 led to the arrest of the Rosenbergs and Greenglass – it was the Rosenbergs who set Greenglass up with the Soviet couriers to whom he was to deliver material from Los Alamos. Greenglass worked on the molds for the explosive lenses for the implosion bomb. Again according to his testimony, he did ask the scientists he worked for about the purpose of the lenses. But he had no scientific training and it is not clear how much he understood of what he was told. At his trial he did produce a sketch that he said was identical to what he turned over to the Russians. It shows a sphere with what looks like the wedge-shaped explosive lenses being put in place. If this was the only thing that the Russians had learned about the Los Alamos project, apart from the very significant fact that there was one, I am not sure that it would have done them that much good. Unlike the next two spies I am going to describe, Greenglass did it in part for the money.

Theodore Hall, nee Holtzberg, was born in New York in 1925, the son of a furrier who came on hard times during the Depression.[5] He had a brother, Ed, who was eleven years older and was essentially the younger brother's surrogate father. Ed went to City College and

---

[5] For an excellent account of Hall see *Bombshell*, by Joseph Albright and Marcia Kunstel, Times Books, New York, 1997. Hall's widow wrote an interesting, albeit self-serving, obituary of him.

was deeply influenced by the Depression. He had two degrees and could not get a job, some of which he attributed to anti-Semitism. He decided to change his last name to "Hall" and over the objections of their parents so did Ted. By his high school years Ted Hall was interested in physics and mathematics. He entered Queens College, which was close to home. When the war began his brother joined the air force and on his leaves he encouraged Ted to leave Queens College for Harvard, where he thought he could get a better education. Although he applied late in the spring of 1942, Ted was accepted for the fall semester as a junior. He was sixteen. By chance, one of his first roommates was a young man named Jack Bean who was the campus chairman of the John Reed Society, named after the 1910 Harvard graduate who wrote *Ten Days That Shook the World*, and had joined the Communist Party. His other roommate, Barney Emmart, was also a member and Ted joined as well. None of them seemed on the surface to take it all that seriously, although Hall did more than the others and even tried to develop his own version of Marxist analysis. In his senior year he roomed with the future Nobelist Roy Glauber, who was as surprised as everyone else when it was revealed that Hall was a spy. Hall and Glauber had had to go through a security clearance procedure and nothing in Hall's background had set off any alarm bells.

When they had completed their course work, having been recommended by a couple of their professors, they were recruited for Los Alamos. Glauber and Hall were both eighteen and Hall, a month younger, was the youngest member of the technical staff. There were two other Harvard recruits, Frederic de Hoffmann and Ken Case. When Glauber and Hall got to Santa Fe, they shared an automobile ride to the mesa with a man who introduced himself as "Mr. Newman." When Glauber signed the registration book he noticed that Mr. Newman had signed in as John von Neumann. The other three Harvard recruits were dyed-in-the-wool budding theoreticians. Only Hall expressed any willingness to work in an experimental group. He worked under the direction of the distinguished cosmic ray physicist Bruno Rossi. Rossi was using a technique that had been invented by Serber called the "ra-la" method for diagnosing implosion trials. For doing this it would be ideal to have some method of recording the progress of the implosion from inside the sphere. Serber had the ingenious idea that if you planted a gamma-ray source in the middle of the sphere and measured how the gammas were transmitted, then the varying densities produced by the irregularities in the implosion would selectively impede the flow of the gammas. You would then be able to take, in a manner of speaking, a gamma-ray photo of the implosion in real time. It was called "ra-la" because the source of the gamma rays that was used was radiolanthanum. Hall joined this activity. A few months later he went back to New York on vacation. By that time he had decided to pass information about Los Alamos to the Russians. He simply felt that the United States should not have a monopoly on this information and that the Russians were allies. But once you have decided this, how do you go about doing it?

Hall had learned about a company called Amtorg whose nominal function was Soviet-American commerce. Its actual function was a cover for espionage. Its purchasing office was on 28th Street, and Hall went there unannounced. He thought he might meet some Russian businessmen. Instead he found a workman stacking boxes. He tried to explain his mission, but he gave the impression of being either crazy or some kind of FBI plant. But he was put in touch with a Soviet agent named Sergei Kurnakov. He went to Kurnakov's apartment and very quickly Kurnakov realized that in terms of intelligence he had just found a diamond in the rough. As a confidence builder Hall gave Kurnakov a list of Los Alamos scientists. From this the Russians could certainly learn that Los Alamos was the real deal.

Kurnakov had to consult with his superiors. They approved Hall, who was given the code name "Mlad." From that time until the end of the war Hall delivered material to the Russians. The question is how valuable this material was. Hall was never apprehended and never confessed, so there is no list as there was with Fuchs. My guess is that it was important, but nothing like what Fuchs delivered. In the first place, Hall was a very junior member of the technical group. His direct information was probably limited to what he was actually working on. He could tell the Russians about the progress in implosion work, but I doubt that he had any knowledge of the properties of plutonium or what the final design of the bomb was. Glauber told me that, over Groves's strong objections, Oppenheimer had insisted on a weekly seminar open to all the members of the technical staff who had so-called white cards, indicating that they had the right level of clearance. I asked Glauber if detailed technical matters – engineering details – were discussed at these seminars. He said no, not because there was a special secrecy about these, but because they were so boring that it would have put everyone to sleep.

In December 1944 Hall was suddenly inducted into the army. This had nothing to do with his espionage activities but simply with the fact that complaints had grown about married men with children being drafted while single people like Hall, who had scientific training, were being deferred. But after some perfunctory basic training, Hall found himself back at Los Alamos, still in the army but as one of the most disorderly privates anyone had ever seen. This status lasted until June 1946 – Hall remained in Los Alamos for a year after most people had departed – when the army revoked his clearance and transferred him to a nonsensitive job at Oak Ridge. Something must have set off an alarm, but still he was never prosecuted.

There was an interesting touch. After Los Alamos began to disband in the fall of 1945, Oppenheimer recognized that the young people like Hall had been deprived of a couple of years of advanced physics courses. He organized a curriculum of twenty-seven courses taught by some of the best people at Los Alamos – Fermi, Bethe, and the like. Hall signed up for two of them. One of them was a course in hydrodynamics taught by Peierls. Hall became the official note taker. He continued in this capacity when Fuchs gave some lectures. He had occasion to ask Fuchs about various aspects of the course. In 1988, when Fuchs died in East Germany, Hall had been living in England, where he had worked in Cambridge since 1962. He was of course aware of Fuchs's confession, although Fuchs certainly never had any knowledge of Hall's past. The Russians were very careful not to cross their agents. They wanted independent information. One can only imagine Hall's feelings when he learned that his Los Alamos hydrodynamics instructor was a fellow Soviet spy. Hall died in England in 1999. From everything he said right up to the end, he had no regrets about the double role he had played at Los Alamos.

Klaus Fuchs was born on December 29, 1911, in Russelsheim, Germany.[6] His father was a Lutheran minister who became a Quaker and a socialist. He was very active in left-wing workers' causes and all of his children inherited his activism. Fuchs was a brilliant student. He first became a member of the socialist party and then in

---

[6] For a useful biography see *Klaus Fuchs, Atom Spy*, by Robert Chadwell Williams, Harvard University Press, Cambridge, Mass., 1987. Williams does not give the full details of what Fuchs turned over, which have only become available recently.

1932 at the University of Kiel he joined the KPD, the German communist party. He had hesitated to join because he disagreed with some of the Marxist dogma, but he decided that the KPD was the only group that was actively opposing Hitler. Over the next few years every member of the Fuchs family was in one way or another a target of the Gestapo. The ones who did not leave the country were often in custody for periods of time. Fuchs realized that he had to leave if he was going to survive and, in September 1933, he arrived in England with all his worldly possessions in a canvas bag. He had Quaker connections in England, and one of them persuaded Neville Mott, who in 1977 won the Nobel Prize in Physics but at the time was a very young faculty member at the University of Bristol, to take Fuchs on with a stipend. Bethe was then Mott's research assistant and later recalled that Fuchs was brilliant, quiet, and unassuming. Some variant of this description was offered by everyone who had any dealing with Fuchs. He got his Ph.D. with Mott in 1936 and then got a postdoctoral appointment with Max Born in Edinburgh, where Born had ended up after he was forced to leave Germany.

Fuchs never made any secret of the fact that he had been a member of the communist party in Germany and that he still felt himself to be a communist. Mott had left-wing leanings, so it did not pose a problem for him. It did, however, pose a problem with the British authorities. Fuchs could not get his German passport renewed and, when he tried, the Germans told the British that the reason was that Fuchs was a member of the KPD. But the British allowed him to remain in England under his expired passport – at least until 1940. He then had to appear before a tribunal in Edinburgh, which classified him as an enemy alien – he was German – and subject to deportation. In fact he was deported to Canada, where he was interned in a camp for enemy aliens. He was not there long. Early in 1941,

Born managed to get him returned to Edinburgh and to find him a war-related job with Peierls in Birmingham. In fact he lived with the Peierls family. Peierls's voluble wife Genia referred to him as "Penny-in-the-Slot" – a reference to automata that you could only get to do something, in this case talk, if you fed them a penny.

At first, Fuchs was only allowed to work on nonsensitive projects. But soon, despite his background, he was cleared to work on Tube Alloys – the British code name for their nuclear program and, in 1942, he became a British citizen. One of the things he did with Peierls was to review whatever German literature was available on matters of fission and the like. Peierls concluded that because the German nuclear physicists continued to publish using their university affiliations, it was very unlikely that they had a major nuclear weapons program. It would have been very clear to anyone who read the American physics journals at this time that the nuclear physicists had disappeared. In fact the Russians came to this conclusion by reading such journals. Fuchs began transmitting information to the Russians almost from the moment he started working on Tube Alloys. His control was a German-born woman whose maiden name was Ursula Kuczynski. She and her husband had visited Moscow in 1930, and she was recruited as an agent. Later she was given the code name Sonia. After living in various European countries she found her way to England in 1942 and settled in Oxford. Fuchs found Sonia after contacting the Russian embassy in London. Fuchs began coming to Oxford from Birmingham, bringing whatever information he had. This was somewhat limited. He had worked with Peierls on critical mass calculations and on gaseous diffusion separation of isotopes, on which Peierls was the leading expert. But he certainly knew nothing about plutonium and nothing about actual bomb design.

After Los Alamos was founded in the spring of 1943, Chadwick put together a British delegation that included Peierls. In the summer of 1944, Fuchs was asked to join them. A remarkable letter has turned up from Chadwick to Peierls dated July 14, 1944.[7] In it Chadwick says, "I have now had a talk with Fuchs himself. He feels that he has a special contribution to make in England, whereas in Y [code for Los Alamos] he would be one of a number and can make no really significant difference to the work." In short, Fuchs did not want to go. One can only imagine what a difference it would have made to the future of the proliferation of nuclear weapons if he had not gone to Los Alamos. At the time that Chadwick wrote this letter Fuchs was already in New York, where he was working on the Manhattan Project. Sonia had given him his American contact, a man with the code name Raymond, in reality Harry Gold, who had been working in Soviet espionage since 1935. Fuchs was persuaded to go to Y.

To understand the magnitude of what Fuchs did, one must understand the position that he found himself in at Los Alamos. He was a senior physicist and an essential part of Bethe's theoretical division. He had carte blanche to explore any part of the operation of the laboratory. He had a photographic memory and, according to people who knew him then, a nearly total mastery of everything that was going on. In addition, people liked him. Not having any social inclinations himself, he babysat for couples who wanted to go out. No one had any idea that he was a spy. In that role he was perfect. By the time of the Trinity test, Fuchs had transmitted the complete design of the gadget. Until 1992, we really did not know the full

---

[7] See http://nuclearweaponarchive.org/Usa/Med/FuchsLetter.html for the full text.

extent of what he transmitted. But that year the Russians released a letter dated October 1945 addressed to Lavrenty Beria. Stalin had named Beria the head of the nuclear weapons program. It escaped no one's attention that he was also the head of the secret police. I quote some of it here so you can see the detail. It is a blueprint.[8] You will recognize some of the things we discussed in the chapter on the gadget.

1. Initiator

An initiator of the "Urchin" type is used in the bomb. It consists of a hollow beryllium spherule on whose inner surface are wedge-shaped grooves. The planes of all the grooves are parallel to one another. The surface of the grooves is covered with a layer of gold of thickness 0.1 mm and a layer of polonium. Inside this spherule is inserted a solid beryllium spherule whose surface is also covered with a layer of gold and polonium.[9]

Dimensions of the "Urchin"

Outer radius of the hollow beryllium spherule	1.0 cm
Radius of base of wedge-shaped groove	0.40 cm
Radius of apex of wedge-shaped groove	0.609 cm
Radius of the solid beryllium spherule	0.40 cm
Number of wedge-shaped grooves	15
Amount of polonium on the surface of all grooves	30 curies
Amount of polonium on solid spherule	20 curies

[8] For the full document see http://nuclearweaponarchive.org/News/Voprosy2.html. I thank Carey Sublette for calling this site to my attention.

[9] The isotope of polonium that is used here is polonium-210, a strong alpha-particle emitter. On November 1, 2006, it was used to poison the former Russian agent Alexander Litvinenko in London. Polonium-210, which has a half-life of 135 days, is produced in reactors. The Russian reactors produce about one hundred grams a year. The evidence is that this polonium came from one of them. One can draw one's own conclusions.

The hollow spherule is made of two halves, which are made in a nickel-carbonyl atmosphere, as a result of which a nickel coating is formed on the surface of the spherule. This coating prevents or at least inhibits the spontaneous decay of polonium.

The initiator works as follows. The shock, directed towards the center, from the explosion of the outer layer of explosive is transmitted through the aluminum layer and tamper, through the layer of active material onto the surface of the hollow beryllium spherule of the initiator. The resulting stresses fracture this spherule along the planes passing through the apex of the wedge-shaped grooves, thus exposing the beryllium of the hollow spherule to the action of the alpha-particles emerging from the polonium coating on the central spherule of the initiator. This produces a neutron flux. The adjacent surfaces of the grooves collide, as a result of which the "Munroe jet" is generated, which penetrates through the thin layer of polonium and gold into the central spherule, thus putting in contact the polonium on the inner surface of the hollow beryllium spherule with the beryllium of the solid one. This also produces a neutron flux.

The neutron flux produced in the initiator attacks the active material.

2. Active Material

The element plutonium of delta-phase with specific gravity 15.8 is the active material of the atomic bomb. It is made in the shape of a spherical shell consisting of two halves, which just like the outer spherule of the initiator, are compressed in a nickel-carbonyl atmosphere. The outer diameter of the ball is 80–90 mm. The weight of the active material including the initiator is 7.3–10.0 kg. Between the hemispheres is a gasket of corrugated gold of thickness 0.1 mm, which protects against penetration of the initiator by high-speed jets moving along the junction plane of the hemispheres of active material. These jets can prematurely activate the initiator.

In one of the hemispheres, there is an opening of diameter 25 mm, which is used to insert the initiator into the centre of the active material, where it is mounted on a special bracket. After inserting the initiator, the opening is closed with a plug, made also of plutonium.

Of course to make use of this, one needed an entire technological infrastructure and a government that was willing to spend any amount of money. The Russians had both.

Fuchs left Los Alamos in 1946, and when he returned to England he began working on developing nuclear weapons for the British. Fuchs had one idea – no country (above all the United States) should have a monopoly on nuclear weapons. One wonders what he would have thought about the current situation. For something like two years he did not engage in espionage, but eventually he changed his mind and walked into the Russian embassy in London and offered his services again. This was considered very bad spy craft. Who knew who was watching? However, they took Fuchs on anyway, and he was given a new control, Alexander Feklisov. Feklisov, who also ran the Rosenbergs in New York, wrote a fascinating book, *The Man Behind the Rosenbergs*[10] about how he became an agent and how he ran first the Rosenbergs and then Fuchs. Part of the fascination is that reading his book is like looking at the world through a mirror that inverts left and right. Both the Rosenbergs and Fuchs are heroes in it, and the Rosenbergs are martyrs. One thing that Feklisov insists on is that none of them were in it for the money. In fact, Fuchs was outraged when he was offered money. At the end he took some because his older brother Gerhard had tuberculosis and was in a very expensive sanitorium in Davos, which he could not afford. Fuchs was caught because the Americans broke the Russian codes.

[10] *The Man Behind the Rosenbergs*, by Alexander Feklisov, Enigma Books, New York, 2001.

Fuchs's name turned up and the British were notified. On December 21, 1949, he was first interviewed by an MI5 officer named Willam Skardon. On January 13, after a third interview, Fuchs confessed. He was tried and found guilty and sentenced to a maximum of fourteen years in prison. There is every indication that he served his sentence stoically. On June 24, 1959, traveling as "Mr. Strauss," he was driven to Heathrow airport and taken to East Berlin on a Polish airplane. He got married and became the director of the Institute for Nuclear Physics at Rossendorf near Dresden. He never expressed any regrets for what he had done, and when asked he said he had no resentment against the British.

That the Russians had spies in Los Alamos is well known. What is less well known is that they had their own ALSOS mission.[11] As many as forty scientists followed the army in the spring of 1945, first into Austria and then into Germany. They were well briefed. On May 4th they went to the Kaiser Wilhelm Institute of Physics in Berlin. They found a treasure trove of documents comparable to what Goudsmit had found in Strasbourg. Among them were Heisenberg's calculations on reactor design. They found that he had used an overly simple geometry. The Russians helped themselves to equipment and, above all, to uranium, which was probably the most valuable thing that they exported. But they also exported people. I have already mentioned Ardenne. There were two other people they captured that probably very few people have ever heard of: Max Steenbeck and Gernot Zippe. While in captivity, they designed a centrifuge that, as I explain in the next, and last, chapter, is at the heart of some of our present problems with nuclear proliferation.

[11] For a very interesting account see German Scientists and the Soviet Atomic Project, by Pavel Oleynikov, *NPR* (Summer 2000).

12. Proliferation

Figure 36. A. Q. Khan. Mian Khursheed/Reuters/Landov.

First we got the bomb and that was good,
Cause we love peace and motherhood.
Then Russia got the bomb, but that's O.K.,
'Cause the balance of power's maintained that way!
Who's next?

France got the bomb, but don't you grieve,
'Cause they're on our side (I believe).
China got the bomb, but have no fears;
They can't wipe us out for at least five years.
Whose next?

Japan will have its own device,
All digital and half the price.
South Africa wanted two, that's right:
One for the black and one for the white!
Who's next?

Iran is gonna get one too,
Just to use on you know who.
So Israel's getting tense,
Wants one in self-defense.
"The Lord's our shepherd," says the psalm,
But just in case, we better get a bomb!
Who's next?

India ignores the ban,
And therefore so does Pakistan.
We'll try to stay serene and calm,
When Venezuela gets the bomb!
Who's next, who's next, who's next?

– Tom Lehrer (revised by Jeremy Bernstein)[1]

IN THE YEAR 1987, A GROUP OF LOS ALAMOS WEAPONS DESIGN-ers, including Carson Mark and Theodore Taylor, published a much-cited article called "Can Terrorists Build Nuclear Weapons?"[2] They seemed to have in mind the construction of rather sophisticated weapons. This is not really relevant for terrorist groups but applies more closely to rogue nations. Here is some of what they write, with commentary.

Most of the schematic drawings that are available relate to the earliest, most straightforward designs and indicate in principle how to achieve a fission explosion, without, however, providing the details of construction. Since 1945, notable reductions in size and weight, as well as increases in yield, have been realized. Schematic drawings of [nuclear

---

[1] The lyrics can be found in *Too Many Songs by Tom Lehrer, with Not Enough Drawings by Ronald Searle*, Pantheon Books, New York, 1981, p. 157. They differ from what is on the original record, where there is no mention of Japan. The last verse is my own devising. In the original, "Venezuela" was "Alabama." Lehrer, a very old friend, invited me to come up with something more current. Another change that I made was to put South Africa in the past tense. Not only is the apartheid reference dated but South Africa did something that is so far unique. It is not generally realized that it had actually built nuclear weapons – about a half-dozen uranium gun-assembly devices. As far as I know, there is no reliable evidence that these were ever tested. But neither was Little Boy. Starting in 1989 South Africa voluntarily gave up its program and dismantled all its weapons, enrichment facilities, and test sites. For this it deserves much credit.

[2] Can Terrorists Build Nuclear Weapons? by Carson Mark, Theodore Taylor, Eugene Eyster, William Maraman, and Jacob Wechsler, Nuclear Control Institute, 2000; http://www.nci.org/k-m/makeab.htm.

weapons of] an entirely qualitative sort are also available that indicate the nature of some of the principles involved in these improvements.

It is difficult to place Fuchs in this continuum. In the communication to Beria that I cited in the previous chapter there are no drawings. But connected with Fuchs's 1948 handover of the Fuchs–von Neumann patent, there are two drawings that exhibit the construction of a gun-assembly weapon. I believe, from what I have been told about them, that they are more than "schematic."

Merely on the basis of the fact that sophisticated devices are known to be feasible, it cannot be asserted that by stealing only a small amount of fissile material a terrorist would be able to produce a device with a reliable multikiloton yield in such a small size and weight as to be easy to transport and conceal. Such an assertion ignores at least a significant fraction of the problems that weapons laboratories have had to face and resolve over the past forty years. It is relevant to recall that today's impressively tidy weapons came about only at the end of a long series of tests that provided the basis for proceeding further. For some of these steps, full-scale nuclear tests were essential. In retrospect, not every incremental step taken would now seem necessary. Indeed, knowing only that much smaller and lighter weapons are feasible, it is possible at least to imagine going straight from the state of understanding in 1945 to a project to build a greatly improved device. The mere fact of knowing it is possible, even without knowing exactly how, would focus terrorists' attention and efforts.

I have tried in the previous chapters to indicate some of the steps that had to be taken to produce "tidy weapons." At the very least, these cannot be done by people living in a cave. However, this is not really relevant to the situation we actually face with terrorists.

In the body of their paper, the authors give the critical masses for various combinations of fissile material. These are what are called "bare crits," meaning critical masses without the use of tampers or boosting. The numbers are interesting. All assume that the material has been reduced to metallic form and is spherical.

Alpha phase plutonium: 10 kilograms
Delta phase plutonium: 16 kilograms
94% uranium-235 – the rest other isotopes: 52 kilograms
50% uranium-235: 160 kilograms
20% uranium-235: 800 kilograms

They estimate that crude weapons could be made with as little as five or six kilograms of plutonium metal or twenty-five kilograms of highly enriched uranium metal. These might produce weapons in the ten-kiloton range. Another option would be to make a crude bomb using tens of kilograms of uranium or plutonium oxide powder instead of metal. Metal requires much more sophistication to produce. They note that plutonium oxide powder might be seized from a fuel fabrication plant and that, if one went this route, then a uranium gun-assembly weapon might be ruled out because of the amount of material needed. However, the calculus changes if all a terrorist group wanted to do was to produce a very crude nuclear weapon – what a weapons designer would call a fizzle. The key ingredient is the fifty, or so, kilograms of highly enriched uranium in metallic form. Given this, it would not be difficult to design a crude gun-assembly weapon that might not be very efficient but could produce, say, a kiloton explosion. If such a weapon had been used at the World Trade Center, the casualties would have been orders of magnitude greater. In a 2006 article by Peter Zimmerman and Jeffrey G. Lewis in *Foreign Policy* magazine, experts on nuclear weapons argue that it would not be very difficult to make a uranium gun-assembly device using a surplus artillery piece. They estimate that such a device would be about nine feet long and would fit comfortably in a Ryder truck.

There are at least two lessons to be learned from these remarks. The first is that when one reads that country x, y, or z has succeeded in producing so many kilograms of a or b and therefore can make n bombs, one needs to ask what assumptions are being made. What kind of nuclear weapons are being discussed? The amount of fissile material needed varies greatly. The second point, which is almost obvious, is that fuel fabrication plants must be made secure. We must not allow these fissile materials to proliferate. Whereas inspecting cargo containers for radiation at ports is important, much more important is securing the fissile material. This is getting harder and harder as more and more countries are producing it.

In 1987, the year that the report of the Los Alamos authors was published, a meeting took place in Dubai that would have profoundly disturbed the report's authors, had they known about it. On its face, it looked like the kind of Byzantine commercial transaction that one associates with the Middle Eastern bazaars. The "sellers" were an odd assortment of individuals from places like Sri Lanka, Pakistan, and Germany. The "buyers" were representatives of the Iranian government, including one Mohammad Eslami, who is now the head of the Revolutionary Guards.[3] The commodity being sold was nuclear technology – the kind that can be used for making bombs. Among the oddly assorted collection of "merchants" were representatives of a Pakistani metallurgist-turned–bomb maker named Abdul Quadeer Khan. The story of how Khan transformed himself from an obscure scientific functionary to the creator of a network that became a veritable supermarket for nuclear weapons

[3] This information comes from *Shopping for Bombs*, by Gordon Corera, Oxford University Press, New York, 2006.

technology is one of the oddest, and most unfortunate, of the twentieth century.

A. Q. Khan was born into a Muslim family in Bhophal, in what was then British India, in 1936. He emigrated to Pakistan, following his older brothers and a sister, in 1952. His father also emigrated and became headmaster of a school near Karachi. I have not been able to find out a great deal about Khan's early life. He attended high school in Karachi and then studied to be an engineer at the University of Karachi. Someone, I have not been able to find out who, helped him to continue his studies in West Germany and then Holland, where he got a degree in metallurgical engineering from the Technical University in Delft. This was followed by a move to Belgium where, in 1971, he got a Ph.D. in metallurgical engineering from the Catholic University in Leuven. He then decided that he would like to stay in Europe to work. His professor, M. J. Brabers, heard about a job opening in an entity that was called the Physical Dynamics Research Laboratory (FDO) in the small Dutch town of Almelo. Khan moved there in 1972. To explain what happened next, we have to return to the Soviet Union.

As I noted in the last chapter, as the war in Europe was ending, the Russians seized all the German scientists and their equipment that they could get hold of. One of the people they got was a physicist named Max Steenbeck. Steenbeck was born in 1904 in Kiel. After taking his degree in 1927, he became head of the scientific department of the German aircraft company Siemens Schukert. He remained there until 1945. One of the things he did at Siemens was work on gas discharges and, during the war, on magnetic mines. In 1945, he was captured by the Russians and sent, after first being put in a concentration camp in Posnan, to Sochumi in Georgia, at the foot of the Caucuses on the Black Sea, with several other scientists. They all lived in a kind of gilded cage. Into this mix, in the summer of 1946, came Gernot Zippe, rescued from a prisoner-of-war camp. Zippe's provenance is a little more obscure. He was born in Austria around 1910 and had a degree in physics from the Radium Institute in Vienna. He served in the Luftwaffe during the war, doing radar and other airplane research in addition to being a pilot. He was flying planes well into his eighties. He was captured and sent to a prisoner-of-war camp. There was apparently a list available to Ardenne of prisoners of war who had scientific training. He spotted Zippe on the list and had him transferred to Sochumi, where he joined Steenbeck's group.

The generic problem that the Sochumi inmates were instructed to solve was the separation of uranium isotopes. There were different groups with different approaches. The most distinguished scientist there was certainly Gustav Hertz. He had won the Nobel Prize in Physics for 1925. He had Jewish ancestry and was hidden away during the Nazi period by Siemens, where he headed their research division. A number of the Farm Hall detainees said that if Hertz had been allowed to join the *Uranverein*, it would have been a different story. They would have gotten much farther. At Sochumi, Hertz's group worked on gaseous diffusion. Ardenne's group worked on electromagnetic separation, which they had been doing during the war. The work on the centrifuge began when Steenbeck demonstrated that the mammoth centrifuges that the Russians had tried could be replaced by a lightweight and relatively inexpensive model. Zippe joined the effort in 1946. He informed me that prior to this he had never worked on centrifuges. There were about sixty staff members. Steenbeck developed the theoretical background while Zippe ran the experimental part.

To whom we owe the final product is somewhat murky. Zippe refers to it as the "Russian centrifuge" but on the web it is generally called the "Zippe centrifuge." By March 1947, the group had begun to experiment with uranium hexafluoride gas. Steenbeck made a trip to Moscow in March 1948 and was given a deadline of one month to produce separated uranium. Zippe produced a simple model that worked. By 1952, the design was mature enough that it could be industrialized and, indeed, the Russians did industrialize it. In 1956, they let Zippe return to Germany, without any papers, but with the design for what was then the best centrifuge in the world in his head. Soon after returning he went to a conference in Amsterdam and realized that the "Russian centrifuge" was better than anything else at that time. Among the people he met was a Dutch centrifuge designer named Jacob Kistemaker. After talking to Zippe, Kistemaker changed the direction of his work and began designing variants of the Russian machine.

Zippe became a consultant for Degussa, one of the largest chemical companies in Germany. Many German companies behaved badly during the Nazi period but Degussa must be close to the head of the list.[4] Their subsidiary, Auer, sold processed uranium metal to the *Uranverein* for their reactor work. The most dangerous part of the work was done by forced laborers from concentration camps. One of Degussa's subsidiaries, Degesch, produced the Zyklon-B gas that was used in the German extermination camps. During the period from 1939 to 1945, Degussa received at least five metric tons of gold to refine, taken from Jews exterminated in concentration camps, especially from their dental work. After the war, they continued their "business is business" modus operandi. From our perspective, the most remarkable example occurred in 1987, the very year that the Iranians were trying to buy nuclear weapons material from A. Q. Khan. That year, coincidentally or not, the Iraqis contacted Degussa to buy nuclear weapons material. One of the things they wanted

---

[4] For a history of Degussa see *From Cooperation to Complicity*, by Peter Hayes, Cambridge University Press, New York, 2004.

to buy was the Zippe centrifuges. As I will explain, the same kind of centrifuge was being sold by A. Q. Khan to the Iranians. The Degussa representatives made it clear that they did not care if the Iranians were going to use the material to make weapons. That was fine with them, as long as they paid their bills. If one can believe it, Saddam Hussein backed away from the deal because he thought that it might be some kind of trap. The overt complicity of the Degussa people made no sense to him. In 1964, the Germans had formed a state-owned company to develop uranium separation technology. In 1970, the company was privatized, and that year it signed an agreement known as the Treaty of Almelo with similar enterprises in England and Holland. FDO, where Khan was hired, was a subcontractor of the Dutch company Ultra Centrifuge Nederland, which became part of the transnational company URENCO. The Dutch hired Khan because they were receiving documents in German about the centrifuge, and they wanted Khan to translate them into Dutch. Obviously he then had access to the complete set of plans for the Zippe centrifuge. He also had access to the Dutch variants. To understand what happened next we have to return to Pakistan.

The Zippe centrifuge (see Figure 37) can produce as many as ninety thousand revolutions per minute. One of the innovations was to heat the bottom so as to produce countercurrents. The heavier uranium-238 is collected in a downward-moving current at the outside while the lighter uranium-235 moves on an upward current on the inside, where it can be collected. The original centrifuges used aluminum rotors, but aluminum has now been replaced by specialized steels. The details of their construction are closely held.

When British India was partitioned in 1947, two Pakistans were created -- East and West -- separated by one thousand miles. There was trouble almost immediately. East Pakistan had a larger population, but the capital of the country was in Karachi in West

GAS CENTRIFUGE

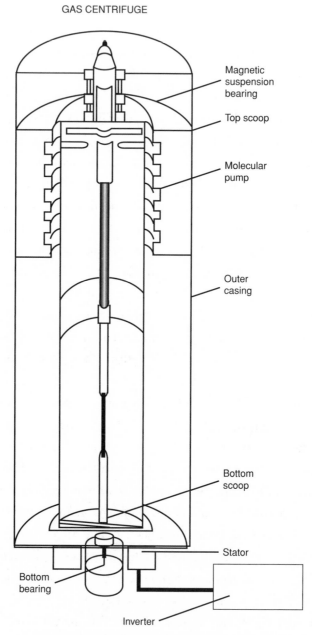

Magnetic
suspension
bearing

Top scoop

Molecular
pump

Outer
casing

Bottom
scoop

Stator

Bottom
bearing

Inverter

Figure 37. A schematic of the Zippe centrifuge showing the flow of gas.

Pakistan. The East Pakistanis, with some justification, felt exploited. Their struggle for independence was, until 1971, kept in check, often brutally, by the Pakistan army. That year, the Indians sent in their own army and Pakistan's army suffered a humiliating defeat. Bangladesh was created. In December 1971, the charismatic and Western-educated Zulifkar Ali Bhutto became president of Pakistan. Even before becoming president, Bhutto had advocated building a nuclear weapon in Pakistan. Once he became president, this was his top priority. As he famously said, they would have such a weapon even if it meant that to finance it, people would have to eat grass. After India tested its weapon in 1974, whatever doubts there were in Pakistan about developing a nuclear weapon vanished. Bhutto tried, with some success, to bring Pakistani scientists back from all over the world to work on it.[5] It was just at this point that A. Q. Khan got into the act. He wrote a letter directly to Bhutto offering his services, which consisted of stealing the plans and parts for the Zippe centrifuge and turning them over to Pakistan. Remarkably, the letter actually got to Bhutto, who approved the idea. Khan was then in business. He decamped from Holland with his Dutch-born wife and set himself up in Pakistan. At first, he was content to work within the official body – the Pakistan Atomic Energy Commission – but he decided that they were moving too slowly. He persuaded Bhutto to give him his own nuclear facility, which he constructed near the modern Pakistani capital of Islamabad. In fairly short order, this entity – the Khan Research Laboratories – was operating according to its own rules with no governmental supervision, although, when it suited

[5] Among the people he persuaded was Munir Khan, who had been with the IAEA and became the head of the Pakistani Atomic Energy Commission. I am grateful to Richard Wilson for this and other information.

everyone, with governmental and especially military collusion. Khan's trump card was that he was the only person in Pakistan, at least in the beginning, who had any idea of how the Zippe centrifuge worked.

Originally, Khan's motives seem to have been entirely patriotic, with a good deal of self-aggrandizement tossed in. He and his representatives scouted Europe to buy the elements needed to make the Zippe centrifuges. Many of the things they needed were dual use, so the real use could be disguised. In most cases, the sellers did not care. In 1979, Bhutto was hanged and a new president, General Zia ul-Haq, took over. He was as determined to get the bomb as Bhutto had been. By this time the Americans had woken up to what seemed like an imminent threat of proliferation. But the struggle to expel the Russians from Afghanistan had begun, and we needed Pakistan's cooperation. A blind eye was turned toward its nuclear activities.

Khan's first extracurricular transaction was with the Chinese. The Chinese had begun their nuclear weapons program in the mid-1950s with Russian help. Among other things, the Russians supplied the design for the implosion bomb that they had originally gotten from Fuchs. Relations soured in 1960, and the Chinese went it alone. They did something very ingenious. Instead of making a plutonium implosion bomb they made a uranium implosion bomb. This saved them the whole step of trying to first make reactors. Indeed, our satellites were looking for Chinese reactors. Not finding any, we concluded, incorrectly, that either the Chinese had hidden them or they had given up the notion of building a bomb. Their 1964 test was a complete surprise, although our satellites did find their separation facility within a few days. But the Chinese lacked advanced centrifuges to enrich large quantities of uranium. Sometime in the early 1980s Khan and the Chinese brokered a deal. In return for the centrifuges, the Chinese would help design the Pakistani bomb. Apart from anything else, they did not want to see India have a nuclear monopoly in Asia. As far as I can see, Khan did not financially profit from this transaction, but his prestige grew exponentially. He then began to run the export of nuclear weapons technology as a business. He opened an office in Dubai operated by his nephew. They soon produced a kind of menu from which you could order, complete with prices. The Iranians bought centrifuge designs and parts of actual centrifuges for several million dollars, which they should have declared to the IAEA. The centrifuge that the Iranians claim to have used to enrich uranium is called the P-1, where "P" stands for "Pakistan." The International Atomic Energy Agency inspectors found, in 2006, bits of Iranian uranium that had been enriched to 40 percent. The only plausible explanation for this is that they had used centrifuges composed of parts of actual centrifuges that the Pakistanis had used to enrich their own uranium. It is very unlikely that the Iranians had then enriched very much uranium – probably only grams – to more than a few percent. In 1998, Pakistan tested its first nuclear weapon. It used the highly enriched uranium produced by the three thousand centrifuges that were operating in Khan's facility in Kahuta near Islamabad. Since 1998, there has been a heavy-water reactor built with Chinese assistance that is producing both plutonium and tritium. In 2006, the Pakistanis announced that they are expanding this program in response to an announced Indian buildup, abetted by America's treaty with the Indians to supply nuclear material.

This treaty, which passed the House and Senate in December 2006, rewards the Indians for what is said to be good behavior on the nuclear front. To many people this agreement seems like a violation of the spirit, if not the actual provisions, of the so-called Treaty on the Non-Proliferation of Nuclear Weapons, often referred to as the Nuclear Non-Proliferation Treaty. This was opened for signature

in 1968 and, at the time of the writing of this book, 188 countries
have signed it, including the United States, Russia, China, and Iran.
Among the nonsigners are India, Pakistan, and Israel. North Korea
signed it and then withdrew. Iran signed it but has threatened to
withdraw. This treaty makes a distinction between nuclear weapons
states and non–nuclear weapons states. Five states – France, China,
the United States, the United Kingdom, and Russia – are classified
as nuclear weapons states. They have agreed not to use their bombs
against nonnuclear states unless they are attacked. The nuclear
weapons states also agree to work toward reducing their stockpiles
of atomic weapons. The so-called Third Pillar is the portion of the
treaty that has caused the most difficulty. This allows the non–
nuclear weapons states to peacefully use nuclear energy and for the
nuclear weapons states to help them. The problem here, as the situation in Iraq showed some years ago and as the situation in Iran
shows now, is that the difference between peaceful and nonpeaceful
uses of nuclear energy is difficult in practice to make. In the absence
of on-the-ground nuclear inspectors it is almost impossible until a
state actually tests a weapon. Two cases in point are Iran and Iraq.

In the 1960s the Russians provided a relatively small reactor to
the Iraqis. In the 1970s the French helped the Iraqis in the effort to
construct a larger reactor. This was to be an intermediate-size "light-
water" reactor. Light water is simply ordinary water whose molecules
consist of two hydrogen atoms and an oxygen atom. In a heavy-
water reactor the hydrogen would be replaced by heavy hydrogen.
In both cases the water is used to slow down the neutrons expelled in
a fission and thus enhance the probability of producing a subsequent
fission. The problem with the light water is the reaction

$$n + p \rightarrow d + \gamma.$$

In words: a proton captures a neutron to produce a deuteron and an energetic gamma ray. This reaction takes neutrons out of the fuel cycle. The advantage with the light water is that you do not need to produce it. You can take it out of a river. To compensate for the loss of neutrons, you enhance the fission probability by enriching the uranium – making it rich in the fissile isotope uranium-235. The French supplied about 27.5 pounds of 93 percent enriched uranium. This is a very high enrichment for the kind of light water reactor used for power production, which can typically operate with something like 3.5 percent enrichment. Using a thick neutron reflector around the uranium sphere made of, say, beryllium, 27.5 pounds of such very highly enriched uranium is almost enough to make a bomb. The uranium was under international safeguards that the Iraqis could have skirted around by withdrawing from the non-proliferation treaty, presumably tipping their hand. This very highly enriched uranium is not very suitable for making plutonium, which converts the isotope uranium-238 into plutonium-239. But if the Iraqis had been allowed to run the reactor, it would have produced depleted uranium – depleted of uranium-235 – which would have been suitable for making plutonium. The Iraqis also had a large supply of their own uranium that they could have used to manufacture plutonium. At least this is what the Israelis assumed when they bombed the reactor on June 7, 1981. It is not clear that they were right. They seemed to equate this reactor with their own Dimona reactor, which was certainly designed to manufacture plutonium. It was after this bombing that the Iraqis decided to enrich their own uranium using Zippe-type centrifuges and to really pursue a nuclear weapon. They paid one million dollars to a German group for the design. They also had calutrons of their own design. The return of the international inspectors after the first Gulf war with a mandate

to search for undeclared nuclear activity showed that they had gotten uncomfortably close.[6]

To use Churchillian language, when it comes to its nuclear program Iran is a "riddle wrapped in a mystery inside an enigma." This
is because the inspectors assigned to verify Iran's activities have not
been allowed to do complete inspections. The inspectors are shown
tantalizing bits with no opportunity to verify anything. For example, the Iranians have made all sorts of claims about their centrifuge
program, which no outsider has been allowed to see. Diplomats who
do not have any scientific background have been shown the interi-
ors of buildings that contain vats and gauges that could mean anything. Although recently, inspectors were allowed to make a surprise
inspection of the centrifuge facility at Natanz and were surprised by
how far the Iranians had gotten. The Iranians insist that the non-
proliferation treaty entitles them to carry out peaceful nuclear development, which they say they are doing. It is not clear why they need
the elaborate centrifuge program they are developing to do this. It
is even more suspicious that they have refused the offer of enriched
uranium from Russia, probably because there would be restrictions
on the extent of the enrichment. There are two reactors under construction in Iran. There is a light-water reactor being built with Russian help at the sea port of Brushehr. This would be fueled with
enriched uranium, and so far the Russians have been dragging their
feet about finishing it and supplying the uranium. This reactor would
not be ideal for plutonium production but it could make do. There is
also a heavy-water reactor being built at Arak that is more suitable
for plutonium production. Incidentally, the Indians got a heavy-
water reactor from the Canadians, who did not get an iron-clad
agreement from them not to use the plutonium for making bombs.

[6] I am grateful to Norman Dombey, Carey Sublette, and Peter Zimmerman for
helpful discussions on this.

Speaking of the Indians...Here the problem, as many see it, is that a country that has refused to sign the non-proliferation treaty is to become a partner in nuclear activities because of its alleged good behavior, as decided unilaterally by the United States. One of the aspects of this partnership is that the Indians can continue with their program of developing nuclear weapons with no inspections. Eight of its reactors will be devoted to such things as the production of plutonium and probably tritium with no constraints while fourteen will be civilian and subject to inspections. The Iranians and the North Koreans, to say nothing of the Pakistanis, may well wonder why they should not get the same kind of deal.

Back to the Pakistanis...In 1988, and again in 1993, Bhutto's eldest daughter, Benazir, was elected president. She was determined to get a missile program to go along with the nascent nuclear weapons development. She wanted long-range missiles. An obvious source was North Korea, which had developed the Nodong missile that had a range of at least six hundred miles. The deal was to pay the North Koreans in cash installments that would total about three billion dollars. Like many shoppers, Benazir found that she was running out of cash before all the payments were made. The solution was barter – missiles for centrifuges – and Khan, who seems to have been dealing with the North Koreans even before this episode, consummated the transaction.

His greatest coup, and his undoing, however, was with the Libyans. For many years Colonel Gadafi had been vacillating about nuclear weapons. In 1995, he made the decision to go ahead with nuclear weapons, and his representatives contacted the Khan network. Khan himself went to Tripoli several times. Ultimately Gadafi bought the whole package for a sum that is thought to be a couple of hundred million dollars. This included the centrifuges, hexafluoride gas, and the plans to make a nuclear weapon using enriched uranium. In 1997, the network delivered twenty P-1 centrifuges, enough to get the Libyans started. But two problems arose. The first was that the network did not have enough material on the shelf to fully supply the Libyans. They had to manufacture some, so they set up what looked like dual-use factories in places like Malaysia. The second problem, which turned out to be fatal, was that the network was penetrated by moles who were never identified. Naïvely one might have thought that this would have brought down the network immediately. The problem was getting the Pakistanis to do it. After the first nuclear test, Khan was practically deified in Pakistan. Moreover, the United States needed the Pakistanis. When the Russians occupied Afghanistan we needed them to allow passage of the Muhajadin into Afghanistan. After the Russians left, we needed them to help in the fight against the Muhajadin we had created. Something dramatic was needed to break the impasse. In October 2003, it happened. A German-owned ship named the BBC China had been engaged to bring the fabricated material from Malaysia. By this time, every movement of the network was known, including this one. When the ship made a stop in Italy it was boarded and the cargo seized. This was the smoking gun that could not be ignored.

Gadafi already had had second thoughts about his nuclear program and the seizure of the BBC China was the last straw. He decided that giving up the program and offering to turn over what he had to the United States would be a wonderful bargaining chip toward diplomatic recognition. There then commenced a series of negotiations that resembled something out of a Monty Python sketch. They lasted until December 11th and, just as the plane with the CIA team and the booty they had gathered was about to leave Tripoli, some Libyans appeared with a stack of envelopes. They turned out to contain the blueprints for the fission bomb that Khan had sold the Libyans. The details of exactly what kind of a bomb this was have not been released. This was something that Pervez Musharraf, who was then the president of Pakistan, could not ignore. On January 31, 2004, Khan was relieved of his position as special advisor to the president. On February 1 Khan confessed to Musharraf what he had done. In his recent autobiographical memoir, *In the Line of Fire*,[7] Musharraf tells us that the first time he knew of the activities of Khan was on September 4, 2003, when he was briefed in Washington by George Tenet, the then director of the CIA, who showed him confiscated plans of the P-1 centrifuge. Musharraf conveniently does not tell us from whom the plans were confiscated. Shortly thereafter, the *BBC China* was stopped, everything became public, and Musharraf felt obliged to begin an investigation. He insists that the government had no role in transferring nuclear material to North Korea. But he does not really deny that such transfers took place. Three days after his confession Khan gave a speech to the nation in English, which made it less accessible to the average Pakistani, apologizing. For many Pakistanis he still was, and still is, a hero – the father of the Pakistani atomic bomb. He was never tried. He was put under house arrest in the rather sumptuous house he built on Rawal Lake near Islamabad. He has been incommunicado. There is a report that he is suffering from prostate cancer. Around the world his network was rolled up – or so one would like to think. In Europe there have been a few trials of his associates that have led nowhere. Some are still living the high life, enjoying the very large sums of money they earned proffering nuclear weapons technology.

Where are we now? Following is a table of nuclear states and their estimated number of warheads.

[7] *In the Line of Fire*, by Pervez Musharraf, Free Press, New York, 2006.

Country	Number of warheads: active/total[a]	Year (name) of first test
Declared nuclear weapons states		
United States	5,735/9,960[b]	1945 ("Trinity")
Russia (formerly the Soviet Union)	5,830/16,000[c]	1949 ("RDS-1")
United Kingdom	< 200[d]	1952 ("Hurricane")
France	350[e]	1960 ("Gerboise Bleue")
People's Republic of China	130[f]	1964 ("596")
India	75–115[g]	1974 ("Smiling Buddha")
Pakistan	65–90[h]	1998 ("Chagai-I")
North Korea	0–10[i]	Not applicable[j]
Undeclared nuclear weapons states		
Israel	75–200[k]	Not applicable or 1979 Vela observation

[a] All numbers are estimates from the Natural Research Defense Council, published in the *Bulletin of the Atomic Scientists*, unless other references are given. If differences between active and total stockpile are known, they are given as two figures separated by a forward slash. If no specifics are known, only one figure is given. If a substantial number of warheads are scheduled for but have not yet gone through dismantlement, the stockpile number may not include all these warheads; not all "active" warheads are deployed at any given time.

[b] U. S. Nuclear Forces, 2006, by Robert S. Norris and Hans M. Kristensen, *Bull. of the Atomic Scientists*, 61, 1, 68–71 (2005).

[c] Russian Nuclear Forces, 2006, by Robert S. Norris and Hans M. Kristensen, *Bull. of the Atomic Scientists*, 62, 2, 64–67 (2006).

[d] British Nuclear Forces, 2005, by Robert S. Norris and Hans M. Kristensen, *Bull. of the Atomic Scientists*, 61, 6, 77–79 (2005).

[e] French Nuclear Forces, 2005, by Robert S. Norris and Hans M. Kristensen, *Bull. of the Atomic Scientists*, 61, 4, 73–75 (2005).

[f] Chinese Nuclear Forces, 2006, by Robert S. Norris and Hans M. Kristensen, *Bull. of the Atomic Scientists*, 62, 3, 60–63 (2006). The Ambiguous Arsenal, by Jeffery Lewis, *Bull. of the Atomic Scientists*, 61, 3, 52–59 (2005).

[g] India's Nuclear Forces, 2005, by Robert S. Norris and Hans M. Kristensen, *Bull. of the Atomic Scientists*, 61, 5, 73–75 (2005).

[h] Pakistan's Nuclear Forces, 2001, by Robert S. Norris and Hans M. Kristensen, *Bull. of the Atomic Scientists*, 58, 1, 70–71 (2002).

[i] North Korea's Nuclear Program, 2005, by Robert S. Norris and Hans M. Kristensen, *Bull. of the Atomic Scientists*, 61, 3, 64–67 (2005).

[j] globalsecurity.org. *Nuclear Weapons Testing – North Korean Statements*.

[k] Israeli Nuclear Forces, 2002, by Robert S. Norris, William Arkin, Hans M. Kristensen, and Joshua Handler, *Bull. of the Atomic Scientists*, 58, 5, 73–75 (2002).

Source: List of states with nuclear weapons, Wikipedia, the Free Encyclopedia, http://en.wikipedia.org/wiki/List_of_countries_with_nuclear_weapons#_note-1#'note-1.

Most of the table is both disturbing and self-explanatory. The Israel entry needs comment, especially the reference to "Vela." The Israeli program is nearly as old as the state. Ben-Gurion authorized it in 1952. The Israelis soon developed new methods of producing heavy water and uranium hexafluoride gas. They interested the French in the project and Israeli nuclear physicists began visits to French weapons laboratories. Using a French design and French assistance the Israelis built a large reactor in the Negev for producing plutonium. When de Gaulle was elected president of France in 1959 he demanded assurances from Ben-Gurion that the reactor was going to be used only for peaceful purposes. Ben-Gurion gave him such assurances, knowing full well the reactor was going to be used to make plutonium. The Israelis began a long game of cat and mouse with the United States, giving ambiguous declarations of intent and voiding real inspections. It might have come to an end with President Kennedy, who was determined to demand inspections, but after his assassination the pressure came off. However, the whole cover was blown by the defection of Mordechai Vanunu in 1985. Vanunu was a Moroccan Jew who, after a brief crash course in the essentials of nuclear weapons production, was in 1977 given the job of supervising the graveyard shift – 11:30 P.M. to 8:00 A.M. – of the reactor. He was able to go anywhere and to take pictures. He was fired in 1985 for his political activities and went to London, where he sold his story and the pictures. It was published on October 5, 1986, in the Sunday *Times*. Vanunu was lured to Rome by a young woman who turned out to be a Mossad agent. He spent many years in solitary confinement in Israel, where he still lives under constant watch.

The reference in the report to "Vela" is also interesting. The Velas were a class of spy satellites that had been equipped with light meters that could respond to millisecond exposures. As I explained before, atomic explosions have a unique visual signature of two very bright flashes separated by milliseconds. The Velas had a practically flawless record of detecting aboveground tests. On September 22, 1979, one of the Velas detected this sort of event. It seemed to be coming from a spot in the south Atlantic Ocean near Prince Edward Island. Naturally, the first assumption was that it was a nuclear test. One guess was that it was an Israeli test, possibly in collaboration with South Africa, which for a while had a nuclear weapons program. One problem was that there was no corroborating evidence – no fallout, for example. For me the most serious problem with the nuclear bomb explanation is that no one has ever come forward to confirm the test. I hope that I have made clear from my description of what I saw in Nevada how complicated the infrastructure of such tests were. They involved dozens of people, from the people who transported the weapon to be tested, to the people who built the towers, to say nothing of the scientists involved. If the Vela observation was a nuclear test, where are the people?

Speaking of warheads, in 2002 President Bush signed a treaty with the Russians to limit the number of deployed warheads to 2,200 per side. This treaty has been much criticized because once it goes into effect on the last day of the year 2012, it could already have been declared invalid because, in the three months prior, each side has the right to withdraw. Even with the treaty, thousands of "operational" warheads are allowed and these can be made ready for deployment very rapidly. Moreover, the United States government has suggested a very extensive modernization program for the existing warheads that will cost billions of dollars and, many argue, gives the wrong signal about proliferation. Advocates for the program claim that the existing "pits," some of which are fifty years old, may have already deteriorated or will deteriorate in the future. Because there has been no testing of these pits with a full nuclear explosion since the signing of the test ban treaty, the advocates of this program argue that we cannot be sure of their reliability. On the other hand, a panel of highly qualified experts studied this matter for JASON, the California-based think tank that works on such issues. In November 2006 they issued a report that said that on the basis of computer simulations and subcritical tests, the pits should remain viable for decades. They saw no need for any modernization program, especially one that would cost billions of dollars. People who believe that such a program is important sometimes argue that not only are the pits aging, but so are the weapons lab designers who could modernize them. They say that these people need to be kept active, otherwise we will lose them. I leave it to the reader to decide on the force of this argument.[8]

When one considers matters of proliferation one must inevitably discuss plutonium. Here matters are more complex than one would at first think. The Nagasaki weapon used what is called "super-weapons-grade" plutonium, with less than 1 percent of any other isotope than plutonium-239. This was because the plutonium was rushed out of the production reactors before the other isotopes could accumulate. Weapons-grade plutonium – as opposed to super-weapons-grade – can contain no more than 7 percent of plutonium-240. Other isotopes are there in lesser concentrations. Reactor-grade plutonium is a mixture of isotopes. If a typical light-water reactor is allowed to run for some time, the plutonium produced is a mixture of about 40 percent plutonium-239, 30 percent plutonium-240, and 30 percent other isotopes. In 1977 it was revealed that in 1962 a bomb made of reactor-grade plutonium had been successfully tested. Neither the precise isotopic mixture nor the yield was revealed, except that it was less than twenty kilotons.

This obviously complicates the proliferation problem. How much plutonium is there? At the end of 2004 it was estimated that globally there were about 1,740 metric tons of nonmilitary plutonium,

---

[8] A good discussion can be found in Future of Nuclear Weapons, by John Dawson, *Physics Today*, February 2007, 24–26.

which grows at about 70 tons a year as the reactors continue to run. The United States has the most, followed by France, which makes most of its electricity using nuclear power, then Japan and Russia. The Russians claim that they have collected most of the civil plutonium from places like Ukraine and Belarus. One hopes so. There has been a long history of fissile material disappearing from poorly guarded facilities.[9] Recently there was a survey of the largest Russian nuclear weapons facility. Young scientists earn about $100 a month. More than 20 percent said that they would sell their services to any state or organization that would pay considerably more. The major nuclear powers have stopped producing plutonium for military purposes. They have all that they can ever possibly use. But the lesser ones such as North Korea, India, and Pakistan are still producing. And then there is Iran. This case is a perfect example of the dilemma that arises when there is a lack of real intelligence. As we mentioned, with Russian help the Iranians are constructing a light-water reactor and, independently, a heavy-water reactor and a facility for manufacturing heavy water. And they have enriched some uranium. Exactly how much, no one knows. Nor does anyone know their real intentions, but it would be foolish to assume that they are peaceful. And, of course there is North Korea.

On October 9, 2006, the North Koreans announced that they had carried out an underground nuclear test. The first indication that this was true came from seismic signals. An underground nuclear explosion produces an artificial earthquake, which sends out seismic signals that differ in their pattern from a real earthquake. A real earthquake is usually preceded by tremors that build up. In a bomb

---

[9] See, for example, *Nuclear Terrorism*, by Graham Allison, Times Books, New York, 2004, for a description.

detonation, nothing precedes the tremor caused by the explosion. It was agreed that the North Korean signals were from a nuclear test. But what kind? To see how this might be determined, recall that the first plutonium fission detected by Hahn and Strassmann produced barium and krypton. Krypton, as an inert gas, floats off. But this is just one of many possibilities for this kind of fission. Two kinds of isotopes of xenon – another inert gas – are produced – xenon-133 and xenon-135. They are produced in a ratio that is characteristic of plutonium fission. An underground nuclear explosion creates cracks in the rocks and this gas can seep through these cracks. In a test like that of the North Koreans this gas propagates for miles and can be detected in trace amounts. Whereas the United States government has not confirmed that this was the method that was used, they did announce that the North Korean test was that of a plutonium device. There is also general agreement that the yield of this device was less than a kiloton – perhaps 500 tons. This is puzzling because the North Koreans had told the Chinese that they were going to test a four-kiloton device. There are two schools of thought about this discrepancy. One school says that it was deliberate. The North Koreans were testing a device that was small enough to be put into one of their missiles. The other school, to which I belong, says that it was a fizzle. The cause of this fizzle, if that is what it was, was not in the plutonium they used. It is certain from the international inspections that took place prior to 1994, when the North Koreans stopped the inspections, that they had produced enough weapons-grade plutonium for a four-kiloton device. Currently, the reactor at Yongbyon is producing enough plutonium for one bomb a year. But I hope that I have made clear in the previous chapters just how complex making a plutonium bomb is. A great many technologies are involved and it would be astonishing to me if on the first try the North Koreans got everything right, although they apparently have mastered plutonium metal production.[10] Thus far the North Koreans have been completely unforthcoming about their uranium enrichment program – the one for which A. Q. Khan was the enabler. Some steps have been made to induce the North Koreans to give up their program, but, as I write this, no one knows how this will end. I keep thinking back to what Stanley Kubrick said to me all those years ago about nuclear weapons being less interesting to most people than even city government. In this age of terrorism, city government cannot be entirely separated from nuclear weapons, and now we all have an interest and a need to know.

[10] The American weapons expert Siegfried Hecker went with a group to North Korea after the test. He wrote a very valuable report for the Center for International Security and Cooperation at Stanford University, from which I have drawn this conclusion.

SUGGESTIONS FOR FURTHER READING

On the general subject of this book there is an embarrassment of riches when it comes to reading material. A good place to start is the two important volumes of Richard Rhodes – *The Making of the Atomic Bomb*[1] and *Dark Sun: The Making of the Hydrogen Bomb*.[2] These books cover a very broad canvas, so choices had to be made regarding coverage. In the atomic bomb book the emphasis is on the more glamorous figures in theoretical physics such as Bethe, Feynman, Teller, and, of course, Oppenheimer. Nowhere will you find, for example, the word "allotrope," which readers of this book will have discovered was key to the use of plutonium in nuclear weapons. A reader who wants to learn more about this fascinating element can try my book, *Plutonium: A History of the World's Most Dangerous Element*,[3] which is written at about the same technical level as this one. For a reader with some technical background I very strongly recommend *Critical Assembly*, edited by Lillian Hoddeson and colleagues.[4] In writing about the hydrogen bomb we are all faced with

---

[1] Simon and Schuster, New York, 1986.
[2] Simon and Schuster, New York, 1995.
[3] Joseph Henry Press, Washington, D.C., 2007.
[4] Cambridge University Press, New York, 1993.

the fact that much of the material is classified. I do not think that it is possible at this time to present a complete history, although Rhodes makes a valiant attempt. Since his book was written, new information has come to light, some of it from Russia. I very much recommend an article by the Russian physicist G. A. Goncharov entitled (in English) "The Extraordinarily Beautiful Physical Principle of Thermonuclear Charge Design."[5] Goncharov was present at the 1955 test of the first Russian hydrogen bomb. This is the first article I have seen that clarifies the role of Klaus Fuchs in this enterprise. He contributed more than what is often attributed to him.

A reader who wants to learn more about Fuchs can try *Klaus Fuchs, Atom Spy*, by Robert Chadwell Williams.[6] For a completely different viewpoint see *The Man Behind the Rosenbergs*, by Alexander Feklisov.[7] Feklisov was the Russian agent who ran both the Rosenbergs and Fuchs. His claim is that Fuchs was considered part of the Russian "team" that made the hydrogen bomb. Much less is known about Ted Hall, one of the other Russian spies. For a fascinating biography see *Bombshell*, by Joseph Albright and Marcia Kunstel.[8] Hall was never caught and never confessed, except to these authors. The whole matter of nuclear spying is discussed in *Spying on the Bomb*, by Jeffrey T. Richelson.[9] For an excellent discussion of the decision by the United States to inaugurate a crash program to make the hydrogen bomb and what might have happened

[5] *Physics-Uspeki*, 48, 11, 1187–1196 (2005). An English translation is available from the British Library.
[6] Harvard University Press, Cambridge, Mass., 1987.
[7] Enigma Books, New York, 2003.
[8] Times Books, New York, 1997.
[9] W. W. Norton, New York, 2006.

if we didn't, see *The Advisors*, by Herbert York.[10] He was involved with all the principals.

Whereas the physics of the hydrogen bomb is still hidden behind a veil of classification, the physics of the fission bomb has been largely revealed. For a reader with a technical background the first source I would suggest is *The Los Alamos Primer*, by Robert Serber.[11] These are the write-ups of the introductory lectures that Serber gave to new recruits to Los Alamos in the spring of 1943. The project was just beginning, so much of what we now know to be important, for example, plutonium weapons, is not in these lectures. But the basic principles are.

Serber was a delightful man and if you want to read more about him I would suggest *Peace & War*, which he wrote with Robert P. Crease.[12] On the other theorists at Los Alamos you might try my *Prophet of Energy: Hans Bethe*.[13] On Oppenheimer there is again an embarrassment of riches. I have written a relatively brief profile titled *Oppenheimer: Portrait of an Enigma*.[14] If you want a more complete biography I recommend *American Prometheus*, by Kai Bird and Martin J. Sherwin.[15] Otto Frisch wrote an excellent autobiography called *What Little I Remember*.[16] In it he tells us about the discovery of the mechanism of fission that he made with his aunt Lise Meitner. The standard biography of Meitner is *Lise Meitner: A Life in Physics*, by Ruth Lewin Sime.[17]

[10] Stanford University Press, Stanford, Calif., 1989.
[11] University of California Press, Berkeley, 1992.
[12] Columbia University Press, New York, 1998.
[13] E. P. Dutton, New York, 1981.
[14] Ivan R. Dee, Chicago, 2004.
[15] Alfred A. Knopf, New York, 2004.
[16] Cambridge University Press, Cambridge, U.K., 1979.
[17] University of California Press, Berkeley, 1997.

If you want to learn about the German war-time program the place to start is *German National Socialism and the Quest for Nuclear Power*, by Mark Walker.[18] My *Hitler's Uranium Club*[19] contains the annotated transcripts of the German physicists' conversations at Farm Hall. Two books on proliferation you might try are *Nuclear Terrorism*, by Graham Allison,[20] and *Shopping for Bombs*, by Gordon Corera.[21]

In this day and age one cannot overlook the web. I very strongly recommend the websites of Carey Sublette, a computer scientist who has made a lifetime avocation of studying nuclear weapons. Try, for example, www.nuclearweaponsarchive.org/Nwfaq/Nfaq4-1.html. I often select a topic that interests me, type it into an Internet search engine, and add Sublette's name. For example, type "implosion" and add "+ Carey Sublette." This usually does the trick.

This sample of books will give you plenty to do, and each of them contains many references for further reading.

[18] Cambridge University Press, New York, 1989.
[19] Copernicus Books, New York, 2001.
[20] Times Books, New York, 2004.
[21] Oxford University Press, New York, 2006.

INDEX

and plutonium production, 279
underground nuclear test, 279–280
withdrawal from NPT, 6
NPT. *See* Nuclear Non-Proliferation Treaty
    (NPT)
nuclear bomb
    Argus project test, 176
    efficiency comparisons, 125
    first explosion of, 152–153
Nuclear Non-Proliferation Treaty (NPT), 5,
    267–268, 270
nuclear reactor
    construction of, 112
    first-run event/December 2, 1942, 112–114
    Hanford B reactor, 115, 142
    "homogeneous" reactor idea, 111–113
    idea for, 109–110
    plutonium producing reactor construction,
        114
nuclear technology, selling of, 259, 262–263
nuclear weapons, by country, 274
nucleus. *See also* compound nucleus
    alpha-particle probes, 33–34
    compound of uranium-238, 57
    of helium, 71, 199
    historical background, 13–20
    liquid-drop model, 47–49, 56
    neutron particle probes, 34
    quantum theory's application to, 79
    structure of, 56–57
    Weizsäcker's mass determination formula,
        49–50

Oklahoma City bombing, 4–5, 185
Oliphant, Mark
    employment of Frisch, 75
    visit to Briggs, 88
"On the Construction of a 'Super Bomb'; based
    on a Nuclear Chain Reaction in
    Uranium" report (Frisch/Peierls), 80–87

On Thermonuclear War (Kahn), 176
Operation Plumbbob, 169, 170
Oppenheimer, Robert
    association with Lawrence, 96

code name "Gadget" creation, 143
Cowpuncher Committee creation, 151
directorship of Institute for Advanced Study,
    138
Los Alamos mesa discovery, 120
Trinity explosion compared to Bhagavad
    Gita verse, 165
oxygen, neutron-proton nuclear equality, 71

Pakistan
    nuclear testing, 5, 267
    number of nuclear weapons, 274
    and plutonium production, 279
Parsons, William, 130
particles, negatively charged, 13, 14, 15
Pash, Boris, 229
Pauli, Wolfgang, 7, 79, 191
Pearl Harbor, 104, 121
Pegram, George, 75
Peierls, Genia (wife of Rudolph), 79
Peierls, Rudolph. *See also* "Memorandum on
        the Properties of a Radioactive 'Super
        Bomb'" report; "On the Construction of
        a 'Super Bomb'; based on a Nuclear
        Chain Reaction in Uranium" report
    background information, 78–79
    creation of theory of solids, 79
    critical mass estimations, 82–87
    professorship at Birmingham University, 79
Penney, William, 137
periodic table of elements
    Mendeleev's development of, 93
    noble gases' location, 94
    organization of, 94
Perrin, Francis, 78
Physical Review, 55, 101, 111
Placzek, George, 56
Planck, Max, 61
"plum pudding" atomic model, 16, 23
plutonium
    allotropic forms, 108
    and alpha particles, 144
    crystallographic density measurement of,
        107–109
    dropped on Nagasaki, 137